ARTICLE 4.

THE FEEDING HABITS OF PSEUDOMYRMINE AND OTHER ANTS.[1]

(Plates I–V.)

By W. M. WHEELER and I. W. BAILEY.[2]

(Read April 24, 1920.)

INTRODUCTION.

All observers of ants have been impressed with their greed and the peculiarities of their feeding habits, and although much has been written on the nature of their food, our knowledge is largely confined to generalities. We know that some species are predominantly insectivorous, some granivorous, some fungivorous and that many others feed to a considerable extent on the saccharine excreta of Aphids, Coccids and Membracids. We also know that certain ants have such specialized appetites that they confine themselves to one or a very few food-substances. Thus some (*e.g.*, *Megaponera fœtens* Fabr., *Carebara*, *Aëromyrma*, *Pœdalgus*, *Liomyrmex*, etc.) feed mainly or exclusively on termites, others on land isopods, or slaters (*e.g.*, *Lobopelta* species according to Wheeler (1904), Arnold (1915) and Williams (1919)), others, like the Attii on particular species of fungi, while still others, intimately associated with some of the higher plants (*Cecropia*, certain neotropical acacias), have long been supposed to eat only particular plant structures or exudates (Beltian and Müllerian bodies, nectar). We are, nevertheless, far from possessing a knowledge of the precise nature, variety and quantity of the food substances ingested by any species of ant, and our knowledge of the substances which these insects feed to their larvæ is even more unsatisfactory. Such a knowledge, however, is urgently needed on account of its bearing on many matters of both theoretical and economic importance.

The feeding habits of animals are so fundamental and have such profound and far-reaching effects not only on their bodily structure but also on their general

---

[1] Contributions from the Entomological Laboratory of the Bussey Institution, Harvard University. No. 170.

[2] There are peculiar rows of excavations in the hollow stems of a number of Ethiopian myrmecophytes that are closely correlated in distribution with the phyllotaxy of the plants. The junior author was able to demonstrate by means of a detailed study of the minute anatomy of the myrmecodomatia that these excavations contain zoöcecidia, which in certain cases are as highly differentiated histologically as many of the complex gall-structures induced by Cynipidæ. In endeavoring to prove that the ants actually feed upon the "nutritive layers" of these zoöcecidia, he had occasion to analyze the castings or pellets from the infrabuccal pockets of a number of the ants. The contents of these pellets were so significant from both the botanical and zoölogical points of view, that it seemed desirable for an entomologist and a botanist to collaborate in studying the feeding habits of various groups of Formicidæ.

physiology and behavior that naturalists have always taken pains to ascertain and record the methods of feeding and the character of the food.   On the other hand we know that the injuries or benefits accruing to the human race from insects are the direct outcome of their feeding habits, so that the economic is even more interested than the theoretical biologist in a knowledge of these peculiarities.   One reason why the economic importance of many common ants remains so dubious or ambiguous is the lack of precise information in regard to the quality and quantity of their food. Moreover, ants are of economic importance not only because they actually feed on certain substances valuable to man, but also because they have a habit of collecting and carrying certain living organisms for considerable distances, either purposely or incidentally.   Sernander (1906), in a beautiful paper, has shown how many of the common European ants—and the same is undoubtedly true of many North American species—are an important factor in the distribution of the herbaceous flora, owing to their habit of collecting and transporting seeds, on the arilli of which they feed. And, as will be shown in the sequel, the extraordinary number and variety of fungus spores and Bacteria which they carry on their bodies or in their mouths, may have a great but hitherto only vaguely apprehended importance for the student of plant, animal and human pathology.

There are at least ten sources that may yield indications of the nature of the food eaten by a particular species of ant.   These are:

1. The prey or booty (living or dead insects, seeds, nectar) carried into the nest.

2. The materials placed on the refuse-heaps, or kitchen-middens, either in the superficial chambers of the nest or on the surface of the soil around the nest entrance.

3. The substances produced by myrmecophiles, larval or adult ants (exudates or glandular secretions) within the nest or by trophobionts (Aphids, Coccids, Psyllids, Membracids, Lycaenid larvæ, etc.) outside the nest.

4. The contents of the pellet contained in the infrabuccal cavity of the adult ants.

5. The contents of the pellet placed in the trophothylax of the larvæ of certain ants (Pseudomyrminæ).

6. The contents of the larval stomach.

7. The meconial pellet of the adult larva.

8. The fungi that may grow on the walls of the nest-chambers or on various substances in the nest (carton, vegetable débris, insect excrement, etc.).

9. Ectoparasitic mites of the ants themselves may be collected and perhaps partially eaten, to judge from the remains of these organisms sometimes found in infrabuccal pockets.

10. The eggs, larvæ, pupæ or even the adult ants of a colony may be devoured under certain usually abnormal conditions, as, e.g., when other food is unobtainable.

All of these sources present problems of considerable difficulty and may easily lead to fallacious interpretations. Thus ants often carry various solids (seeds, buds, flowers, leaves, bird, mammal and insect excreta, etc.) into the nest for no apparent reason, except to avoid returning home with empty mandibles. Such materials may be later thrown on the kitchen-middens and create the erroneous impression that they have been partially eaten. The exudates and secretions of myrmecophiles and of larval ants are notoriously difficult of observation in particular instances. As will be shown in the sequel, the pellets contained in the infrabuccal pocket and the trophothylax consist in large part of extraneous substances derived from cleaning the surfaces of the body or from excavation in friable soil or wood, and the contents of the larval stomach and meconial pellets are in most cases so finely triturated and have been so much altered by the action of the digestive juices as to be undeterminable. Furthermore, the meconial pellet can be recovered, as a rule, only in the subfamilies of ants which make cocoons (Ponerinæ and Formicinæ). The cropping by ants of mycelia growing in their nests is by no means proof that the fungus filaments are actually eaten. Cannibalism, though not infrequent, is, of course, an exceptional or abnormal source of food.

On the other hand, since the pellets of the infrabuccal cavity and trophothylax and the materials in the stomach of the larva often contain recognizable fragments of food as well as substances excavated or manipulated by the ants in making their nests, careful analysis, especially of the pellets, often throws considerable light not only on the nature of the food but also on that of the environment in which the ants nest. For this reason we have dissected out and examined several hundred of the pellets in a considerable number of species, representing well-known genera in all the sub-families (except the Cerapachyinæ) and have devoted considerable space to an account of the substances found. Particular attention was devoted to the Pseudomyrminæ, because it is the only subfamily in which it is possible in the same species to study both the pellets of the infrabuccal pocket of the adult and those of the larval trophothylax.

The whole subject is thus seen to be one of considerable scope and complexity. Our results have been obtained exclusively from a study of preserved material and must therefore be regarded as merely furnishing a foundation for future experimental researches on the living ants. Such researches will be most advantageously pursued in tropical or subtropical regions where the family Formicidæ is represented by the greatest number of genera and species of morphological and ethological interest. Though the scope of our work is circumscribed, it yields, nevertheless, some indications of the magnitude and importance of a study of the trophic activities of ants and may furnish some useful hints to future investigators.

I. THE FEEDING HABITS OF ADULT ANTS.

A. HISTORICAL.

An adequate account of feeding in adult, or imaginal ants would require a description of the structure and functions of the entire alimentary tract. This would be a work of supererogation, because these subjects have been set forth in sufficient detail in several treatises, such as those of Donisthorpe (1915), Escherich (1906) and the senior author (1910). There is one structure, however, the infrabuccal pocket, which may be considered more exhaustively, as well as certain peculiarities of feeding in such specialized ants as the harvesters and fungus-growers, since in these cases it seems to be most difficult to determine the precise composition of the substances actually taken into the crop and stomach and assimilated by the insects.

The Dutch entomologist Brants (1841), while studying the mouth-parts of various Hymenoptera, discovered in all three phases of wasps (*Vespa crabro*, *vulgaris* and *holsatica* and in *Odynerus*), but not in bees, Ichneumonids and Crabronids, a peculiar subspherical pouch just behind the tongue and below the buccal orifice. He discussed the structure at length and supposed at first that it might be used either as a receptacle in which the materials for the paper nest could be prepared, or as a crop for the feeding of the young. He accepted the former as the more probable function and therefore called the structure the "lijmholte." On examining it under the microscope he found it to contain small fragments of wood .1 to .5 mm. in length, in which he clearly recognized the cells with their pitted walls. He also found algæ, but supposed that they had been growing on the wood, and though he detected some animal substances, such as a piece of spider's skin, he believed them to be accidental and that the pouch was properly a mortar in which minute wood-particles could be mixed with saliva and thus made into paper.

The same pocket was discovered in ants by Meinert (1860), but called the "Mundsak." He regarded it as being morphologically an infolding of the hypopharynx and says that he often found it filled with a "dark brown, formless mass." The walls were described as clear and transparent and devoid of musculature. He dissented from Brants' opinion in regard to its function, pointing out that it is present in male and female as well as worker ants, that it is well-developed in at least the majority of European species, none of which, except *Lasius fuliginosus*, makes a carton nest, and that in this species the organ shows no enlargement or peculiarities of structure. He was inclined to believe that it may function, at least to some extent, as a kind of crop for feeding the larvæ or adult ants. He therefore returns to an hypothesis which Brants suggested but rejected.

Lubbock in 1877 gave a very similar account of the organ, with figures illustrating its appearance in median sagittal sections of the head of a female *Lasius niger* and a worker *Myrmica ruginodis*. Concerning the structure, for which he adopts Meinert's term "mouth sac," he makes the following statements: "Its membranous walls appear to be firm and elastic. I have not found any muscles attached to it, and presume that it is kept open by the elasticity of the walls. The orifice of the sac can be closed at will by a small flap (Pl. CLXXXIX, Fig. 3 *fl.*) which is supplied with several muscular fascicles (Plate CLXXXIX, Fig. 3 *M M*). The cavity generally contains a brown, spongy mass, in which I once (in a specimen of *Formica rufa*) found a small hematoid [*recte* Nematode] worm." Lubbock also discusses the mechanism employed in emptying the sac. He finds Brants' account of the process obscure and concludes that in ants "the general action of the muscles which open the pharynx would tend to empty the mouth sac; those which draw down the lower wall of the pharynx, by directly constricting the mouth sac, while even those attached to the upper wall of the pharynx would tend to empty the mouth sac by sucking out its contents." His figures and his description of the pharynx are also of interest. "The walls of the pharynx are more or less chitinous, and the lower portion immediately above the entrance to the mouth sac is covered with small teeth, which point downwards. This arrangement is unusual, since in most animals the palatal teeth point backwards, so as to prevent any food which has once entered from returning; and, on the contrary, to promote its passage down the throat. I presume, however, that in ants, which feed principally on animal and vegetable juices, it is advantageous to prevent the entrance of solid particles."

Adlerz (1886) considers the "mouth sac" rather briefly, referring to the results of Brants and Meinert and differing from the latter in regard to its function. He believes that the crop (ingluvies) and not the "mouth sac" serves to carry the food to the larvæ, and adds that in *Camponotus*, *Lasius fuliginosus* and other ants which commonly nest in decayed wood, "the mouth sac is often found filled with wood-fragments and earth particles," in *L. fuliginosus* "mixed with a dark brown humor," which arises, not from the cells of the organ itself, but from some of the glands in its neighborhood. Adlerz makes no statement as to what he considers the true function of the mouth sac to be.

Janet was the first accurately to describe the method of feeding in adult ants and social wasps. His results were presented in several papers (1895a, 1895b, 1897, 1900, 1905). He found the sac described by Brants, Meinert, Lubbock and Adlerz just behind the upper surface of the labium and beneath the buccal orifice, as a small subspherical pouch and called it the "cavité supralabial," "poche prébuccal," "cavité de moulage des boulettes," or "poche gnathale" (Text-fig. 1 h).   At various times the senior author has referred to it as the "infrabuccal cavity, or chamber" or "hypopharyngeal cavity."   Donisthorpe (1915) has adopted the term "infrabuccal

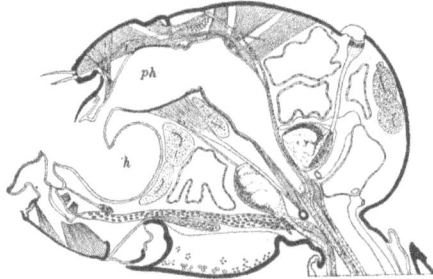

FIG. 1. Sagittal Section of head of femur *Lasius niger* L. (After C. Janet) *ph*, pharynx, *h*. infrabuccal pocket.

cavity" in his monograph of the British ants, Escherich (1906) employs the term "Infrabuccaltasche" and Stitz (1914) "Intrabuccaltasche," possibly a typographical error for "Infrabuccaltasche."   We prefer to designate it as the "infrabuccal pocket" throughout our discussion.

Janet showed that whereas liquid food is directly imbibed by ants from the surface of the labium and carried into the pharynx and thence through the tenuous œsophagus into the crop, particles of solid food, obtained by licking with the tongue or rasped off by the maxillæ, are carried into the infrabuccal pocket where they are moulded in the form of a pellet, which he calls a "corpuscle enroulé," or "corpuscle" or "boulette de nettoyage."   He demonstrated that this pellet, which as we have seen, had been noticed by Meinert, Lubbock and Adlerz, consists not only of bits of food but also in large part or even exclusively of particles of foreign matter which have been cleaned off from the surfaces of the body by the tongue or the strigils, the comb-like organs—really modified spurs—at the tips of the fore tibiæ (Plate I, Fig. 2).   The strigils, used primarily and most beautifully adapted for cleaning the antennæ, are frequently drawn over the maxillæ and labium, thus depositing the particles that adhere to the teeth of their combs.   The particles are probably kept together by the secretion of a gland which Janet found to open by many minute pores at the bases of the strigillar teeth.   But ants also use the tongue in cleansing one another and their brood and hence many particles of foreign matter may be carried more directly into the infrabuccal pocket and agglutinated or mixed with the food particles of the pellet.   The solid portions when no longer of nutritive value, are cast out as a small pellet of refuse.

Janet studied the pellets of wasps and of several species of ants and by powdering the insects with various substances demonstrated the fact that the pellets contain particles from the surface of the body. He says of the conditions in the ants: "The composition of the 'boulettes de nettoyage' shows that they arise from the food residue and the dust and detritus which the ants remove by means of their toilet organs from their own bodies, from their progeny and from their companions. Sometimes two ants may be seen facing each other, in a position that might lead one to suppose that one of them, with its mandibles wide open, is regurgitating liquid food to the other, whose mandibles are closed. Actually, however, they may be engaged in a different occupation, the worker with the closed mandibles being absorbed in carefully licking and cleaning the open mandibles of its companion. This may be easily observed in large species such as *Camponotus ligniperdus*. It is mainly the mandibles which are subjected to such thorough cleaning, but it extends also to the other buccal appendages and neighboring parts of the head, parts which the animal possessing them doubtless cannot cleanse as easily and completely as can one of its companions." Concerning the conditions in the hornets Janet makes this statement 1895*b*): "The hornets, like the other wasps and the ants, not infrequently cast out little rolled-up corpuscles from the supralabial cavity. These corpuscles are usually 1.2 to 1.5 mm. long and 1 to 1.2 mm. broad. All powdery substances accidentally deposited on the hornets' bodies and the solid residue of their meals are found thus molded in the form of rolled-up corpuscles. The chitinous fragments of malaxated insects and the wood-paste that serves in the construction of the nest, commonly figure among the elements of these corpuscles, which in shape are in all respects comparable with the corpuscles I have already described in the ants."

That the analysis of the pellet in the infrabuccal cavity is capable of yielding very interesting and important results was demonstrated by H. von Ihering (1898) and Huber (1905) in their studies on the fungus-growing ants of the neotropical tribe Attiini. These authors showed that the peculiar fungus, which the large Brazilian *Atta sexdens* L. cultivates on a substratum of triturated vegetable substances such as pieces of leaves or caterpillar excrement, is transmitted generation after generation, from the maternal to the daughter colonies, in the form of the food-pellet of the recently fecundated *Atta* queen. Before leaving the maternal nest for her marriage flight she fills her infrabuccal pocket with a mass of hyphæ from the fungus gardens. After excavating her own small cell in the ground and closing the aperture, so that she is entirely shut off from the outside world, she spits out the pellet. The hyphæ at once begin to grow in the moist atmosphere, deriving their nutriment from the detritus, or cleanings contained in the pellet, and later the queen adds her own feces and crushed eggs as manure, thus keeping the diminutive garden alive and flourishing till her first batch of small workers have hatched.   They then break through the soil and collect and bring in the vegetable substances required as a nutrient substratum for the continuous growth of the fungus.

Recently Bruch, in a preliminary note (1919), has given a similar account of colony formation by an allied fungus-growing ant, *Acromyrmex lundi* Guérin of Argentina.   He says, however: "The female ant, before leaving the formicary, takes into her buccal cavity a pellet consisting of vegetable substances containing fungus spores.   After the nuptial flight she loses her wings, enters the soil and makes a cavity, the initial chamber of the future fungus-garden.   Twenty-four hours later she deposits the pellet, the spores of which germinate in the humid environment, producing slender filaments, the fungus hyphæ."   It thus appears that, in this species at least, some of the vegetable substratum of the fungus-gardens of the parental nest is carried over in the infrabuccal pocket, but since true spores are not produced by the fungus *Rhozites gongylophora*, according to Moeller, it is rather difficult to understand how the gardens of the daughter colony can arise in the manner described by Bruch. His completed account, when published, will no doubt clear up this difficulty.

The feeding habits of the harvesting ants have been rather carefully investigated by Moggridge (1873), McCook (1879), Neger (1910) and Emery (1899, 1912). Moggridge studied *Messor barbarus* and its subspecies *structor* at Mentone, and confirmed many statements of the ancient Greek and Roman authors who had seen these ants harvesting, storing and eating the seeds of various plants.   He ascertained that the seeds may be kept for a long time in a sound condition in the chambers ("granaries") of the nests, without showing signs of germination.   When germination does occur in the nests—and Moggridge believed that the ants usually soften the seeds and make them sprout—"before they are consumed by the ants, it is very curious to see how the growth is checked in its earliest stage, and how, after the radicle or fibril—the first growing root of dicotyledonous and monocotyledonous seeds—has been gnawed off, they are brought out from the nest and placed in the sun to dry, and then, after a sufficient exposure, carried below into the nest."   And he adds: "The seeds are thus in effect malted, the starch being changed into sugar, and I have myself witnessed the avidity with which the contents of seeds thus treated are devoured by the ants."   On Plate 6 he shows how the ants mutilate the germinating seeds by cutting off the radicle, thus confirming statements of Plutarch and Pliny.   Moggridge does not, however, go into the actual details of feeding in *Messor*.

McCook in his voluminous study of the Texan harvester, *Pogonomyrmex barbatus* subsp. *molefaciens*, gives a much more detailed account.   The seeds are broken open or squeezed by the mandibles and "the juices or oils of the seed, and the fine, starchy powder of the grains rasped off by the tongue. . . . The process of eating," he says, "is a steady licking of the surface of the seed." Although inclined to believe that *Pogonomyrmex* prefers sound seeds as food, he nevertheless cites Mrs. Mary Treat as having "formed the opinion from her observations of the Florida harvesters [*P. badius*], that they eat the grains only after or during sprouting, and that their appetite especially affects the saccharine substance, which is only manifest at fermentation." After discussing the salivary glands of ants he cites Lubbock's observations on the infrabuccal pocket and surmises that the "pappy mass of flour" licked by the ants from the seed, may have first found lodgment within this sac," but at the time he was studying *P. barbatus* he was not aware of the English naturalist's observations.

In 1899 Emery published a very brief but important paper on the feeding of the Italian *Messor barbarus* subsp. *structor* in artificial nests of the Janet pattern. He noticed the ants cropping the mycelium which grew on the walls of the nest and inferred—perhaps erroneously—that the insects were feeding on the fungi. They also fed on green seeds and buds, boiled, dried beef, hulled rice and other ripe grain, whole or in fragments, bread and the wheat paste (semolina) employed by the Italians in making macaroni, spaggetti, etc. Crude starch, however, was constantly rejected. The wheat paste was the principal food used in the experiments. "This paste was given to the ants in small round grains. The ants held these grains between their mandibles during whole days till they swelled and softened to form a ductile dough which could be kneaded. Then the residue was abandoned in the dry, illuminated chamber of the nest. With this aliment alone the ants have reared young larvæ to the perfect state." He therefore believes that the insects extract from this food not only sugar, but also and mainly the proteids. From the fact that they also utilize boiled, dried beef containing very little soluble material he infers that "the saliva of ants seems to be able to peptonize albuminoid substances, as Plateau has proved for *Periplaneta orientalis*. The fact that these ants reject crude starch leads to the belief that they are unable to dissolve it. Be this as it may, we are confronted with a digestion that occurs or at least begins in the mouth, though it may be completed in the crop under the influence of the saliva."

Neger studied the habits of *Messor barbarus* subsp. *meridionalis* on the island of Arbe, off the Dalmatian coast. He found that the ants chew up and insalivate the soft part of the seeds to form small pasty masses which he calls "ant-bread crumbs" ("Ameisenbrodkrümmel"), which are carried out and placed on the surface of the soil to dry. Although he never saw them being removed he supposes that the ants eventually carry them into the nest. These crumbs, he finds, frequently contain an amylolytic and proteolytic fungus (*Aspergillus niger*) which renders them more nutritious so that they can be fed to the young as "larva-bread" ("Larvenbrod").

Emery in 1912 regards Neger's conclusions as "addirittura fantastiche"—downright fantastic, and confirms his own former observations on *Messor structor* by an experimental study of the Italian *Messor barbarus* subsp. *minor* kept in artificial nests. The ants were given their choice between the embryonic and starchy portions of wheat grains and were found to prefer the former. He concludes that this is why the ants attack the germinal or radicle end of the seed and not because they wish to prevent its growth. The cutting off of the radicle, he believes, is due to similar causes though the reasons for this behavior "are, perhaps, more complicated." He agrees with Neger that the germination of the seed is merely useful to the ants in facilitating the removal of the often very hard glumes and coatings. As Neger observed, the process of germination is not sufficiently advanced to produce an abundant transformation of the starch into maltose. Emery also fed his ants on the wheat-paste (semolina) made up in the form of small rings. When given to the insects these were malaxated like the softer parts of the seed and converted into contorted masses corresponding to Neger's ant-bread crumbs. With the assistance of Prof. L. Pesce, Emery determined the weight and starch content of 50 rings before and after their malaxation. The reduction in starch-content in the rejected crumbs varied from 7.3 to 15.51 per cent., but other substances, in all probability proteids, had also been extracted by the ants. He infers that the workers of *M. barbarus minor* either digested or fed to their larvæ at least 7.3 per cent. of the starch in the paste, that they consumed also an unknown quantity of nonamylaceous substances, probably proteids, and that the latter are a much more important aliment than the starch. He concludes with a statement which precisely confirms the observations of the senior author on the various species of *Pogonomyrmex* in the Southwestern States: "The granivorous ants are derived from insectivorous ants and represent an adaptation to the climatic conditions of dry prairies, steppes and deserts. When, owing to the summer droughts, insects become scarce and are no longer sufficiently numerous to satisfy the needs of the ants, the granivorous species substitute the living but dried seeds of plants, but at least the species I have observed, will not refuse any insects that may be obtainable. The seeds, however, have the very great advantage that they can be stored for a long time, that they can be accumulated in granaries and that they contain abundant provisions, not exactly for the winter alone, as the ancients maintained, but for any periods of scarcity."

B. OBSERVATIONS.

We find that the infrabuccal pocket in the numerous Formicidæ we have dissected agrees closely with Janet's account and figures. Among the latter we select his sagittal section of the head of a queen *Lasius niger* L. for reproduction as Text-fig. 1, as it shows the relations of the pocket (*h*), the pharynx (*ph*), buccal orifice, tongue, musculature, and nerve ganglia (brain) of the head very clearly. The tongue is

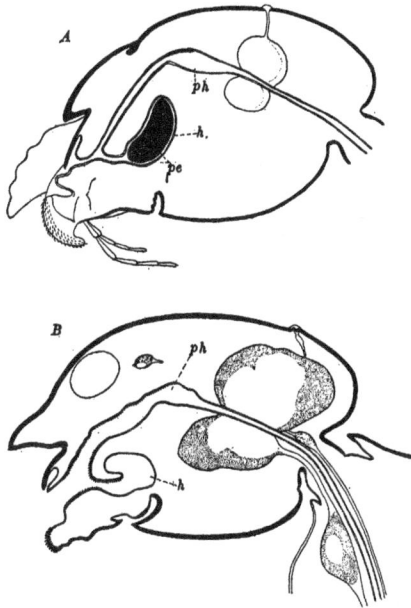

FIG. 2.   *A*, Sagittal section of head of female *Camponotus* (*Myrmoturba*) *brutus* Forel.   *B*, Sagittal section of head of mature worker pupa of *Viticicola tessmanni* Stitz.   *ph*, pharynx; *h*, infrabuccal jacket; *pe*, pellet.

represented as depressed, so that the mouth is wide open. In Text-fig. 2 sagittal sections of the heads of two other ants are represented. *A* is a drawing of the median sagittal surface of the right half of the head of a female *Camponotus* (*Myrmoturba*) *brutus* Forel, with the mouth in the usual closed condition and the pellet (*pe*) *in situ* in the infrabuccal pocket. This view was obtained by simply bisecting the insect's head in the median line with a Gillette razor blade. The pellet stands our very conspicuously among the white surrounding tissues as a blackish, slightly curved, sublenticular body, filling the infrabuccal pocket, which determines its form, and showing a distinct stratification, each layer of which evidently represents the compacted strigil-sweepings of a single toilet operation.

Text-fig. *B* is a sagittal section of the head of a mature worker pupa of *Viticicola tessmanni* Stitz and is introduced, because it shows the very fine rows of chitinous spinules on the tongue, buccal and pharyngeal walls, originally described and figured by Lubbock. It will be noticed that these spinules, as he observed, point forwards in the anterior portion of the pharynx and backwards about the opening of the infrabuccal pocket, so that particles of solid matter would naturally tend to be kept out of the alimentary tract proper and to be directed into the pocket.

Although previous observers have called attention to the thin wall of the pocket and to the complete absence of musculature, no attention seems to have been devoted to the structure of the chitinous lining. We find that the latter has a beautiful pattern of polygonal areas, evidently the expression of the underlying, very flat hypodermal cells and that this pattern appears to vary so much in different species and genera that it might furnish characters of some taxonomic value, at least in certain groups of Formicidæ. We have not, however, deemed it advisable to devote much attention to such details in the structure of the infrabuccal pocket because their bearing on the investigation in hand is too remote.

The pellets, with which we are here concerned, were rather easily dissected out, each was mounted in a drop of glycerine on a slide and then crushed or spread out by pressure on the cover glass. The following table is a very condensed enumeration of the contents of the pellets, 736 in number, taken from species belonging to all the subfamilies, except the Cerapachyinæ and Pseudomyrminæ. No material of the Cerapachyinæ was available, and the Pseudomyrminæ may be reserved for special consideration (pp. 256 and 260).

The following analysis of the infrabuccal pellets of 38 different ants, of very different habits and from widely separated localities reveals several striking results. It will be seen that remains of insect food are by no means as abundant as would be expected. Owing to the nature of such food, however, softer particles can be identified only with difficulty, so that in many instances we have had to place a query in the column in which they are recorded. On the other hand, spores and particles of plant tissue occur in the pellets with surprising constancy, and are often numerous or very abundant, quite irrespective of the nesting habits.

The subfamily Pseudomyrminæ embraces only four genera: *Tetraponera* F. Smith (= *Sima* Roger), *Pachysima* Emery, *Viticicola* Wheeler and *Pseudomyrma* Lund. *Tetraponera* comprises a large number of species which are distributed over Ethiopian Africa, Madagascar, Indomalaya, Papua and Northern Australia. *Pachysima* includes only two species, *aethiops* Fabr. and *latifrons* Emery, and is confined to western equatorial Africa. *Viticicola* is monotypic, the single species, *tessmanni*

PONERINÆ.[1]

Ant.	Locality.	Nesting in	Food.			
			Insect Tissue.	Plant Tissue.	Fungi.	No. Pellets.
Myrmecia tarsata Sm........	N. S. W.	Ground	S	A	F–VN	14
M. gulosa Fabr.............	N. S. W.	Ground	S–A	A	N–VN	8
Paraponera clavata Fabr.....	Panama	Ground	S	A	VN	10
Diacamma australe Fabr.....	Queensland	Ground	S	A	VN	16
Odontomachus hæmatoda L. ...	Costa Rica	Ground	S	A	N–VN	10

DORYLINÆ.

| *Dorylus* (*Anomma*) *nigricans* Illiger var.............. | Congo | Ground | A | ? | F | 18 |
| *Eciton burchelli* Westwood... | British Guiana | Ground | A | A | F–N | 10 |

MYRMICINÆ.

Leptothorax curvispinosus Mayr..................	N. Y.	Dead twigs	?	A	N	41
Aphænogaster tennesseensis Mayr..................	Mo.	Rotten stumps and logs	?	A	N	30
Veramessor pergandei Mayr..	Ariz.	Ground (H)	?	A	F	18
V. andrei Mayr............	Calif.	Ground (H)	?	A	F	12
Pogonomyrmex barbatus Sm. var. *marfensis* Wheeler...	Texas	Ground (H)	S	A	F–N	46
P. barbatus Sm. subsp. *rugosus* Em..............	Ariz.	Ground (H)	S	A	F–N	42
Solenopsis geminata Fabr. var. *rufa* Jerdon........	Philippines	Ground (H)	?	A	N	6
S. geminata var. *diabola* Wheeler..............	Texas	Ground (H)	?	A	F–N	18
Myrmicaria eumenoides Gerst..................	Congo	Ground	?	A	F–VN	12
Atta cephalotes L..........	(3) Costa Rica	Ground	?	A	F–VN	22
Trachymyrmex septentrionalis McCook var. *obscurior* Wheeler..............	Florida	Ground	?	A	F–N	8
Cryptocerus multispinus Em.	Guatemala	Dead branches	S	A	F–VN	8

DOLICHODERINÆ.

Liometopum apiculatum Mayr	Ariz.	Carton nests in hollow live oak	?	A	F–N	10
L. occidentale Em..........	Calif.	Carton nests in hollow live oak	S–A	A	A–VN	18
Leptomyrmex nigriventris Guérin.................	N. S. W.	Hollow logs	S	A	A–VN	14
Dolichoderus (*Hypoclinea*) *doriæ* Em..............	N. S. W.	Ground	S–A	A	F–VN	19
Azteca instabilis Sm........	Guatemala	Carton nest	S	A	VN	24

[1] S = fragments of tissue scanty
A =   "   "   " abundant
F = spores or fragments of hyphæ infrequent
N =   "   "   "   "   " numerous
VN =   "   "   "   "   " forming considerable fraction of total volume of pellet
H = seed storing ants

FORMICINÆ.

Ant.	Locality.	Nesting in	Food.			
			Insect Tissue.	Plant Tissue.	Fungi.	No. Pellets.
Œcophylla longinoda Fabr....	Congo	Silk and leaf nests on trees	S	A	N	16
Lasius (Dendrolasius) fuliginosus Latr...........	Switzerland	Carton nests in hollow trees	?	S	VN	33
L. (Acanthomyops) interjectus Mayr.................	Illinois and Colorado	Ground	?	A	F–N	12
Formica rufa L. var. *obscuripes* Forel.............	B. C.	Mound nests	?	A	F–VN	17
Myrmecocystus melliger Llave	N. Mex.	Ground	?	A	F	12
C. herculeanus L. subsp. *ligniperda* Latr. var. *noveboracensis* Fitch	(2) N. Y.	Decaying stumps	S	A	N	41
Camponotus (Myrmoturba) maculatus Fabr. subsp. *maccooki* Forel var. *sansabeanus* Buck..........	Texas	Ground	A	S	F–N	11
C. (Myrmoturba) nigriceps Sm.....	Queensland	Ground	A	A	F–VN	18
C. (Myrmoturba) acutirostris Wheeler..............	Ariz.	Ground	S–A	A	F	7
C. (Myrmoturba) brutus Forel	(3) Congo	Rotten wood	S	A	VN	50
C. (Myrmothrix) abdominalis Fabr. var. *costaricensis* Forel.................	Costa Rica	Dead trees	S	A	A–VN	17
C. Myrmepomis sericeiventris Guérin...............	Guatemala	Rotten tree trunks	?	A	VN	35
Polyrhachis (Myrmhopla) dives Sm...............	Philippines	Silk and detritus nests on trees	A	A	F–VN	10
P. (Hagiomyrma) ammon Fab.	N. S. W.	Ground	S–A	A	F–VN	23

Stitz having apparently an even narrower range in the Kamerun and Congo Basin. *Pseudomyrma* is exclusively neotropical, extending from Argentina to Texas, Southern California and Florida and represented by a great number of species, especially in Central and northern South America. Our taxonomic knowledge of this genus is unsatisfactory, because Frederick Smith published very poor descriptions of many of the species and because Forel, who more carefully described many others, very rarely figured his types. The difficulties are increased by the fact that many of the species are highly variable and are often represented by inadequate series of specimens in our collections.

All the species of Pseudomyrminæ nest in the cavities of plants, except *Pseudomyrma elegans* Smith, which nests in the ground or in the walls of termitaria (*vide infra*). The very long, slender body of these ants and of their brood is evidently an adaptation to living in narrow cavities and is most pronounced in the female of *Ps. filiformis* Fabr. (See Wheeler 1919). It is apparent also in other ants which live under similar conditions (*Cylindromyrmex, Simopone, Metapone, Podomyrma, Leptothorax,* some *Camponotus,* etc.).

Although the senior author's collection contains a large number of Pseudomyrminæ, we have selected for the following study only the species which were accompanied by well-preserved larvæ. The reasons for this will be evident from the sequel. The list of forms is as follows:

1. *Tetraponera allaborans* Walker (Text-fig. 4D). Los Baños, Philippines (F. X. Williams), from hollow twigs.

2. *Viticicola tessmanni* Stitz (Text-figs. 4E and 6). Medje, Belgian Congo (Lang, Chapin and Bequaert), nesting in the hollow stems of a Verbenaceous liana, *Vitex staudtii.* For an account of the larva see Wheeler, 1918, p. 303, Fig. 6.

3. *Pachysima aethiops* F. Smith (Text-fig. 4B). Several localities in the Belgian Congo (Lang and Chapin), nesting in the hollow twigs of a Rubiaceous myrmecophyte, *Barteria fistulosa.* For an account and figs. of the larvæ of all stages see Wheeler, 1918, p. 305, Figs. 7 and 8.

4. *Pachysima latifrons* Emery (Text-fig. 4C). Niangara, Belgian Congo (Lang and Chapin), nesting in the hollow twigs of *Barteria fistulosa.* See Wheeler 1918, p. 308, Figs. 9, 10, for a description and figures of the various larval stages.

5. *Pseudomyrma gracilis* Fabr. (typical) Text-figs. 3, 4A and 5). Corozal and Las Sabanas, Panama (Wheeler), nesting in hollow twigs; Quirigua, Guatemala (Wheeler), living in hollow thorns of *Acacia.*

6. *Ps. gracilis* var. *mexicana* Roger. Cuernavaca, Mexico; San Jose and Cartago, Costa Rica; Lake Atitlan and Guatemala City, Guatemala (Wheeler), nesting in hollow twigs. This variety occurs as far north as Brownsville and Victoria, Texas.

7. *Ps. gracilis* var. *dimidiata* Roger. Patulul and Escuintla, Guatemala (Wheeler), in hollow twigs.

8. *Ps. gracilis* var. nov. Escuintla, Guatemala (Wheeler), nesting in *Acacia* thorns.

9. *Ps. rufomedia* F. Smith. Antigua and Guatemala City, Guatemala (Wheeler), in hollow twigs. The specimens agree with Smith's very poor description and are from the same locality as the types. The species has not been recognized by recent authors.

10. *Ps. elegans* F. Smith. Kartabo, British Guinana (Alfred Emerson), nesting in a termitarium. According to Forel this is the only *Pseudomyrma* known to nest in the ground.

11. *Ps. belti* Emery. Escuintla, Guatemala (Wheeler), nesting only in the thorns of acacias (*A. cornigera* L.). This and the two following forms feed on the Beltian bodies and nectar of their host-trees.

12. *Ps. belti* var. *fulvescens* Emery. Quirigua and Zacapa, Guatemala (Wheeler), like the type of the species, nesting only in *Acacia* thorns (*A. hindsi* Benth.).

13. *Ps. spinicola* Emery. Las Sabanas, Panama (Wheeler), nesting in the thorns of *Acacia sphaerocephala* S. & C.

14. *Ps. sericea* Mayr var. *fortis* Forel. Escuintla, Guatemala (Wheeler), nesting in the hollow internodes of *Triplaris macombii* (D. Smith).

15. *Ps. championi* Forel var. Lake Atitlan, Guatemala (Wheeler), nesting in hollow twigs.

16. *Ps. filiformis* Fabr. Patulul and Zacapa, Guatemala (Wheeler), nesting in dead branches. The three phases and the habits of this species are described in a recent paper by the senior author (1919).

17. *Ps. decipiens* Forel. Cartago, Costa Rica and Antigua, Guatemala (Wheeler), nesting in hollow twigs.

18. *Ps. caroli* Forel. Guatemala (Wheeler) nesting in hollow twigs.

19. *Ps. elongata* Mayr. Bahamas and Cuba (Wheeler), nesting in hollow twigs.

20. *Ps. flavidula* F. Smith. Andros and New Providence Islands, Bahamas (Wheeler), nesting in hollow twigs and the culms of grasses.

21. *Ps. flavidula* var. *delicatula* Forel. Guadalajara, Mexico (McClendon), Austin, Texas and San Jose, Costa Rica (Wheeler), nesting in hollow twigs.

22. *Ps. arboris-sanctæ* Emery. Frijoles, Panama (Wheeler), in hollow trunks and branches of *Triplaris cummingiana.*

23. *Pseudomyrma* sp. Patulul, Guatemala (Wheeler); in hollow twigs.

24. *Pseudomyrma* sp. Antigua, Guatemala (Wheeler), in hollow twigs.

25. *Pseudomyrma* sp. Escuintla, Guatemala (Wheeler), in hollow twigs.

26. *Pseudomyrma* sp. Cartago, Costa Rica (Wheeler), in hollow twigs.

The pellets were dissected out of several workers of a number of the Pseudomyrminæ in the foregoing list, but as they had precisely the same composition as the pellets taken from the trophothylaces of the cospecific larvæ, apart from a somewhat smaller number of particles of insect tissue, a detailed account of the components is here omitted and the reader is referred to the description of the larval pellets on p. 261.

II. FEEDING HABITS OF LARVAL ANTS.

A. HISTORICAL.

In 1918 the senior author rather hastily reviewed what was known concerning the feeding of various ant-larvæ. It was shown that the most general method was with liquid food regurgitated by the worker nurses, a method probably almost universal in many species of certain subfamilies (Myrmicinæ, Dolichoderinæ and Formicinæ) for the larvæ throughout their development. In this connection attention may be again called to the following observation by Miss Fielde (1901) on the young larvæ of the Myrmicine *Aphaenogaster fulva* Roger: "The feeding of the larva, which is bent nearly double in the egg, with regurgitated food begins as soon as it straightens itself and protrudes its mouth. When the larvæ begin to appear in the egg-packet, the workers lift the packet and hold it free and still, while one of their number holds a transparent white globule of regurgitated food to the larval mouth projecting from the surface of the egg-packet. I have repeatedly seen the workers thus feeding the very young larvæ, a single globule of regurgitated food serving for a meal of which four or five larvæ successively partook."

Newell (1909) studied the feeding of the larva in a Dolichoderine, the Argentine ant, *Iridomyrmex humilis* Mayr, and describes the process as follows: "The larva ordinarily lies upon its side or back. The attending worker approaches from any convenient direction, usually from one side or from the direction in which the head of the larva lies and, spreading her mandibles, places them over the mouth-parts of the larva which are slightly extruded. The tongue of the worker is also in contact with the larval mouth. While the worker holds body and mandibles stationary a drop of light-colored, almost transparent fluid appears upon her tongue. This fluid disappears within the mouth of the larva, but it cannot be ascertained to what extent the larval mouth-parts are moved during the operation, owing to their being obscured from view by the mandibles and head of the attending worker. Slight constrictions of the larval abdomen during feeding are sometimes noticeable, at other times not. The time required for feeding a single larva varies from 3 to 30 seconds, depending doubtless on the hunger of the "baby." The workers proffer food to, or at least inspect, each larva, for the worker doing the feeding will place her mandibles to the mouth of one larva after another, feeding those which seem to require it."

Undoubtedly many Ponerinæ and even some Myrmicinæ and Formicinæ feed their larvæ directly with whole insects or pieces of insects. The facts relating to this method of feeding, especially in the Ponerinæ were also reviewed by the senior author in 1918. Janet has seen the larvæ of the European *Lasius flavus* DeG. feeding in this manner, and the senior author has made similar observations on the American species of *Lasius* belonging to the subgenus *Acanthomyops*, on *Aphaenogaster fulva* and *Pogonomyrmex imberbiculus* Wheeler. Workers of *P. imberbiculus* were given houseflies, and the senior author, writing in 1902, stated that "these were not only eaten with avidity by the adult *Pogonomyrmex*, but cut into pieces and fed to the larvæ in the same manner as I have described for the Ponerinæ and Myrmicinæ (1900). On one occasion nearly every larva in the nest could be seen munching a small piece of house-fly."

Certain agricultural ants, which feed on seeds (*Pogonomyrmex*, *Messor*) also nourish their young with the same food. Thus the workers of *P. imberbiculus*, referred to above, were seen to bring seeds from their granaries, crack them open and "after consuming some of the softer portions themselves, to distribute the remainder among their larvæ. The latter could be seen under the lens, cutting away with their mandibles and devouring the soft, starchy portions of the seeds." Emery (1915, p. 185) has observed the same method of feeding the larvæ in *Messor barbarus* L. He says: "The granivorous ants also nourish their larvæ with seeds, either in the form of the regurgitated contents of the crop, or in the form of seed-fragments saturated with saliva and directly administered.   In an artificial nest I have also seen a worker of the court-yard ant carrying in her mouth one or two larvæ attached to a piece of masticated seed."

Still another type of feeding prevails among the fungus-growing Myrmicinæ of the tribe Attiini.   Tanner in 1892 observed *Atta cephalotes* L. in Trinidad and reported as follows: The eggs become enveloped in a "pearly white fluffy growth," evidently masses of fungus hyphæ.   The larvæ, on hatching from these eggs "are usually placed on top of the nest and are constantly attended by the smallest workers—the nurses—who separate them into divisions according to size.   At first it seemed a mystery, how these minute grubs could be fed so systematically, knowing that each individual larva was only one among so many, yet certain it was, that all were equally attended to.   Further observations showed that nature had provided most efficiently for them to ask for food when they required it.   This the larvæ do by pouting their lips;   at this notification of their requirement the first nurse who happens to be passing stops and feeds them.   The nurses are continually moving about among them with pieces of fungus in their mouths ready for a call for food.   The nurses feed the minute larvæ by merely brushing the fungus across their lips showing that the spores alone are sufficient for its food at that period of its life.   But it is not so when the larvæ have increased so much in size, that the pout can be seen without a glass, for then the whole piece after having been manipulated by the nurse's mandibles into a ball, in the same manner as the leaves are served, when they are first brought into the nest, is placed in its throat and if that is not sufficient the pout continues when the next one and even the next passing proceeds with the feeding, till the pout is withdrawn, showing that it is satisfied.   No further notice is then taken of it by the feeders, until it again asks for a meal by pouting later on in the day."

Tanner's observations were made on adult colonies.   J. Huber (1905), however, shows that the conditions in incipient colonies of the closely allied *Atta sexdens* L. are different.   As soon as the first larvæ hatch, they are fed directly with eggs placed in their mouths by the queen.   To quote part of his description, "after the mother ant has laid an egg she first palpates it for several seconds and then turns to a larva, which she tickles with her antennæ till it begins to move its jaws, whereupon she thrusts the egg, usually with some force, end foremost between the jaws, which continue to move against it.   The egg sometimes stands off perpendicularly from the body of the larva, sometimes and more frequently it lies more or less closely applied to the larva's ventral surface.   In the latter case the mother ant often presses the egg down with her foot.   If the larva is still small, the egg is usually taken away after a short time and given to another larva; a large larva, however, is able to suck out an egg completely in the course of three to five minutes, so that only the collapsed shell remains, which is later licked away by the mother. . . . I believe that eggs, at least till the appearance of the first adult workers, are the exclusive food of both the mother and her brood. I have never seen *Atta* females giving their larvæ the mycelium or kohlrabi of the *Rozites*. And, unlike von Ihering, I have never seen the mother ant devouring kohlrabi." The best proof that the larvæ of the first brood do not receive fungus food is that the *Atta* queen is occasionally able to bring them up without a fungus garden. Later, according to Huber, after the first workers have matured the larvæ are fed by the latter with the peculiarly modified hyphæ called "kohlrabi" by Moeller and Huber.[1]

According to Bruch (1919), the queen of the Argentine fungus-grower, *Acromyrmex lundi*, also nourishes the larvæ of her first worker brood with some of her eggs.

We may now turn to a consideration of the Pseudomyrminæ, the larvæ of which were first described and figured by Emery in 1899. His remarks, referring to *Pseudomyrma* and *Sima* (*Tetraponera*) are here translated *in extenso*: "These two genera present larvæ of a very peculiar type. I have examined those of *Pseudoponera flavidula* F. Sm. (Cayenne, collected by Pillaut) and of certain species of *Sima*, particularly *S. natalensis* F. Sm. and *S. clypeata* Emery (Cape Colony, collected by Dr. Brauns).

"The larvæ are subcylindrical anteriorly and somewhat narrowed behind; the first postcephalic segments are more developed dorsally, shortened ventrally, so that in profile they seem to have a fan-shaped arrangement, their dorsal outlines forming together a curve or convexity, which constitutes the apparent anterior extremity of the larva, while the head, or morphological anterior end is found to be situated on the ventral surface of the body. Hence these larvæ may be called *hypocephalic*, in contradistinction to those belonging to the great majority of ants, which I would call *orthocephalic*.

"In *Sima* (Fig. 7) the head is depressed and scarcely projects from the ventral surface of the larva, its buccal extremity being, when in a state of repose, embedded in a cavity of the third and fourth segments of the trunk on which it rests. In *Pseudomyrma* (Fig. 8), the head is rounded and distinctly projecting and the third and fourth segments of the trunk are not hollowed out to receive it. The maxillæ

---

[1] A recent study by the senior author of various Attiine larvæ, belonging to the genera *Atta*, *Trachymyrmex*, *Cyphomyrmex*, *Sericomyrmex*, and *Apterostigma*, shows that their mouthparts are beautifully adapted for the methods of feeding described by Tanner and Huber. The mandibles are short, stout and acute, and except in *Apterostigma*, covered with numerous sharp spines, so that the delicate egg-shell or the thin walls of hyphal filaments or of kohlrabi spherules can be held and firmly squeezed and at the same time perforated with many small openings, thus allowing the liquid contents to be rapidly expressed and trickle into the mouth.

have no conical projections, there being in their places groups of small tubercles. Two similar groups are seen on the labium, where the cones are also lacking. The mandibles are small, slightly projecting, but robust and bidentate at the apex. A character quite peculiar to the larvæ of *Sima* and *Pseudomyrma* is the presence on the head of a pair of small appendages which I regard as rudiments of antennæ. Each of them consists of two small unequal conical or subcylindrical projections terminating in a very minute, obtuse hair (olfactory hair?). The hairs of the body are short and simple, but one observes, distributed along the body, four double series of long setæ, uncinate at the tip and regularly arranged on the single segments, as shown in Fig. 7*a*." The last remark, on the hooked hairs, refers to *Tetraponera*, as Emery neither figures nor describes the homologous structures in *Pseudomyrma*.

In 1912 Emery gave a brief account and a figure of the full-grown larva of *Pachysima aethiops*. He saw some of the peculiar exudatoria in the final stage of their development but erroneously interpreted them as the "beginnings of legs."

In 1918 the senior author figured and described the various stages of the larvæ of *P. aethiops* and *latifrons* and the larva of *Viticicola tessmanni*, but at that time his attention was mainly concentrated on the development and structure of the exudatoria. Nevertheless the structure of the trophothylax and enclosed food-pellet was figured and briefly described.

The only other published account of immature stages of Pseudomyrminæ is a study of the embryology of an unidentified species of *Pseudomyrma* by Strindberg (1917), treating only of the early stages. The type of cleavage, which occurs in two periods, is unique or unusual, not only in the Formicidæ but among insects in general. The serosa is absent, a peculiarity which the Swedish embryologist has also observed in *Leptothorax* and *Tetramorium*.

B. Observations.

The stomach contents of the great majority of ant-larvæ are so finely comminuted and have been submitted to such protracted digestion that it is impossible to determine their precise nature. It will be shown below how the comminution is probably brought about. Occasionally a minute fragment of chitin or a few hairs may indicate that parts of insects have been ingested. Still there are a few genera that show very clearly that the larvæ are nourished with pellets of insect flesh or with rather coarse fragments of insects. We have found peculiar conditions in the Dorylinæ. Examination of a number of half- and nearly full-grown worker larvæ of *Eciton burchelli* Westwood received from Mr. Wm. Beebe, shows that the stomach is unlike that of other known ant-larvæ in being very long and slender and in having unusually thick, muscular walls. The larva is fed, apparently at considerable intervals, with rather large pellets consisting of the rolled up soft-parts of insects. These pellets are so compact that they retain their form in the narrow lumen of the stomach, where they lie in an irregular longitudinal series. Occasionally minute fragments of chitin or a few fungus spores are present, but owing to the feeble development of the larval mouth-parts so characteristic of the Dorylinæ, it is evident that the worker must prepare these pellets by carefully trimming away the hard, chitinous portions of their insect prey and rolling up the denser, muscular portions of the flesh. The worker probably consumes much of the exuding juices while engaged in this operation and before stuffing the pellets into the gullets of the larvæ.

The other genera in which the larval stomach was found to contain evidence of the precise nature of the food are *Cataulacus*, *Cryptocerus* and *Leptothorax*. We have examined the larvæ of *Cryptocerus egenus* Santschi, *Cryptocerus minutus* Klug, *C. multispinus* Emery, *C. (Cyathocephalus) varians* F. Smith and *wheeleri* Forel and some eight species of *Leptothorax* of the sugbenera *Leptothorax* sens. str., *Mychothorax* and *Goniothorax* from such widely different localities as the Congo, Central America, the Northern United States and Southern Europe. In all the forms cited the larval stomach is voluminous and closely packed with coarse chitinous fragments of small insects (Plate I, Fig. 8), in some cases interspersed with numerous fungus spores (Plate I., Fig. 6). In some species of *Leptothorax* the whole contents consist of entire or nearly entire legs of small insects. The mandibles of the larvæ of these three genera are short, broad and stout and therefore well-adapted to crushing, so that the coarse fragments may have been bitten off by the larvæ from larger pieces or whole insects proffered by their worker nurses. The pieces may, however, have been cut up to a considerable extent by the workers. The fungus spores may have come from their infrabuccal pouches, but some of the coarse insect materials at least, could hardly have such a provenience, especially in such small ants as the species of *Leptothorax*.

An examination of the meconial pellets, taken from the cocoons of the Ponerinæ and Formicinæ, furnishes no satisfactory information in regard to the larval food, apart from the fact that they contain black or dark brown, very finely comminuted solid matter. A biochemical investigation, which we are not competent to undertake, would probably reveal the presence of significant decomposition products.[1] The

---

[1] The workers and soldiers of highly carnivorous ants, such as the Dorylinæ and certain species of *Pheidole*, e.g., *Ph. ecitonodora* Wheeler and *fallax* Mayr, *Megaponera foetens* Fabr. and *Paltothyreus tarsatus* Fabr., have a very powerful odor like that of the species of *Chrysopa* and therefore very similar to indol. Melander and Brues (1906) regarded the substance as probably being a leucin. Though this substance seems to be normally present in the feces of insects, there is some doubt as to whether in *Eciton* it is derived from the alimentary tract or from certain dermal glands, such as those in the epinotum.

Ponerinæ, with very few exceptions, are so patently and exclusively insectivorous that it is unnecessary to resort to the meconium for information concerning their diet. In the Pseudomyrminæ interesting information is obtained from a study of the pellets

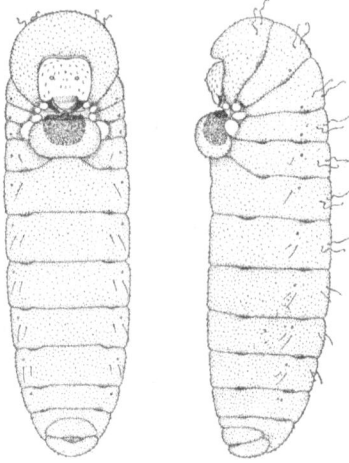

FIG. 3.   Ventral and lateral view of larva of *Pseudomyrma gracilis* Fabr.

found in the trophothylaces. Before entering on this study, however, it will be advisable to introduce a more comprehensive account of the peculiar larvæ of this subfamily.

The adult larvæ of all four genera of Pseudomyrminæ are much alike (Text-figs. 3 and 4). The body is long, straight and cylindrical, not broader posteriorly as in nearly all other ant-larvæ. The anterior and posterior extremities are blunt and rounded and the segments are all sharply defined. The integument is uniformly thin and perfectly transparent, though tough, only the mandibles, as a rule, being strongly chitinized and the lining of the buccal cavity somewhat pigmented. The prothoracic segment is large and hood-shaped, and in certain species can be drawn down over the head; the meso- and metathoracic segments are narrowed ventrally, the head is large, somewhat flattened, usually subrectangular, about as broad as long and embedded in the ventral portions of the thoracic segments. The antennal rudiments are always distinct as small, rounded papillæ, each bearing three sensillæ. The mandibles are small, stout and bidentate, sometimes with a vestige of a third tooth, their upper surfaces covered with regular rows of subimbricate papillæ. The maxillæ are large, swollen and rounded, lobuliform, the labium short and broad, with the transverse, slit-shaped opening of the salivary duct in the middle. The sensory organs which in many other ants have the form of papillæ or pegs on the maxillæ and labium are in the Pseudomyrminæ usually reduced to small areas or feeble eminences, bearing the groups of sensillæ. The anterior maxillary organ has five, the posterior two and each labial organ has five of these sensillæ. The buccal cavity is broad and transverse, its dorsal and ventral walls being in contact and both furnished with fine, regular

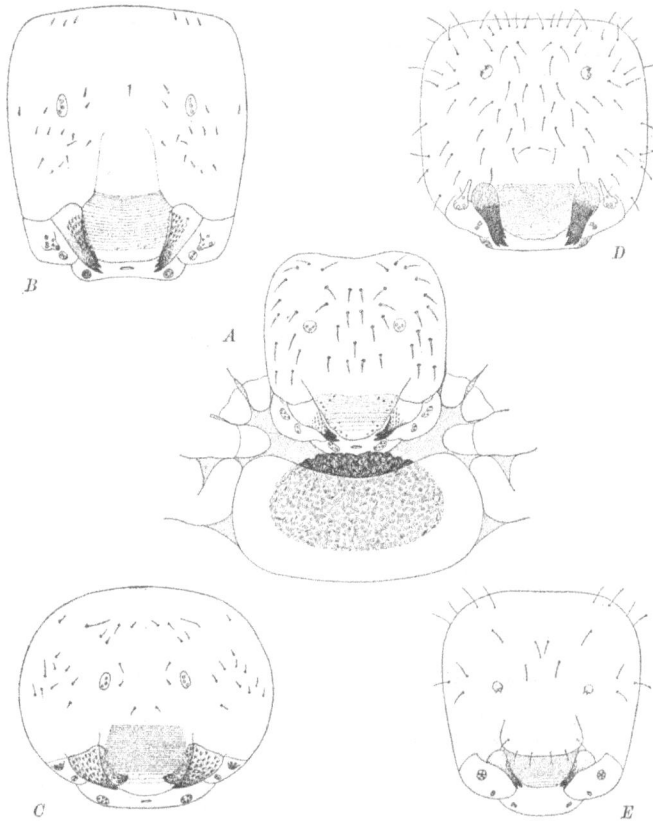

FIG. 4. *A*, Head, trophothylax and exudatoria of larva of *Pseudomyrma gracilis* Fabr. *B*, Head of *Pachysima aethiops* F. larva. *C*, Head of *P. latifrons* Emery larva. *D*, Head of *Tetraponera allaborans* Walker larva. *E*, Head of *Viticicola tessmanni* Stitz larva.

transverse ridges. This peculiar structure, the *trophorhinium*, will be described in greater detail below. Each thoracic segment bears a rounded papilliform exudatorium ventrally on each side next to the head. The sternal portion of the first abdominal segment is transversely elliptical, swollen, protuberant and furnished with a food-pouch, the *trophothylax*, opening forward, *i.e.*, towards the mouth-parts (Text-figs. 3 and 4*A*).

The hairs on the body of the larva are of three kinds: first, short, stiff, very acute hairs, generally and rather evenly distributed over the whole surface (*microchaetæ*); second, much longer, stouter, more gradually tapering, lash-like and somewhat curved hairs of unequal length, singly or in a row or loose cluster on each ventrolateral surface of each abdominal segment (*acrochaetæ*), and third, long hairs, of uniform length, only slightly tapering, with hooked tips (*oncochaetæ*). These are normally present in transverse rows of four to eight on the dorsal surfaces of the three thoracic and first three to eight abdominal segments. On the more posterior segments they are often represented by simple, *i.e.*, pointed hairs.

In the genera *Tetraponera*, *Viticicola* and *Pseudomyrma* the youngest larvæ, apart from their proportionally longer and more conspicuous oncochaetæ and acrochaetæ and more protuberant trophothylax, have essentially the same structure as full-grown individuals. In the two species of *Pachysima*, however, as the senior author has shown (1918), the youngest larvæ are very unusual in possessing long, stout, blunt bristles in the place of the oncochaetæ and extraordinary exudatoria which may have the form of appendages on the three thoracic and first abdominal segments. The following generic and specific modifications of the principal characters detailed in the preceding paragraphs were noticed in the various Pseudomyrmine larvæ examined:

Tetraponera allaborans (Text-fig. 4*D*).—Head rather rounded behind. Anterior maxillary sense-organs produced into slender, anteriorly directed points. Oncochaetæ straight, without sigmoidal flexure, in four rows, a pair in each row, on each segment from the prothoracic to the sixth abdominal. Similar hairs, but without hooks, occur on the seventh to ninth abdominal segments. Acrochaetæ absent. Microchaetæ short and sparse, much longer on the head and somewhat longer on the prothorax than on the more posterior segments.

Viticicola tessmanni (Text-fig. 4*E*).—Larva very slender, with very prominent trophothylax and large rounded exudatoria on each side of it. Oncochaetæ straight, four in number, in two pairs on each segment from the prothoracic to the sixth to eighth abdominal. Acrochaetæ very long but unequal, numerous, on the side of each abdominal segment. Microchaetæ short, unequal, scattered.

Pachysima aethiops (Text-fig. 4*B*).—In large larvæ there seem to be no traces of oncochaetæ or acrochætæ. Microchaetæ short, of unequal length. Head rather subrectangular, mandibles broader than in the three other genera and the sense-organs of the maxillæ appearing as distinct though low tubercles. Labrum trapezoidal, its anterior border truncate, entire.

Pachysima latifrons (Text-fig. 4*C*).—In large larvæ the oncochaetæ are present, but very short, stout, curved or feebly sigmoidal, eight on each segment from the metathoracic to the seventh abdominal. Acrochaetæ absent; microchaetæ much longer than in *aethiops*. Head transversely elliptical. Mandibles broader and flatter, with broad, blunt apical tooth. Papillæ bearing the maxillary sensillæ less prominent. Labrum with more rounded anterior border.

Pseudomyrma (Text-figs. 3 and 4*A*).—Head distinctly subrectangular, as long as broad. Exudatoria present as rounded papillæ on the ventrolateral borders of the three thoracic and first abdominal segments, and much alike in the youngest and oldest larvæ. Trophothylax more protuberant in the youngest stages, but throughout larval life with essentially the same structure as in the three other genera. Mandibles small, with 2–3 strongly chitinized, rather blunt teeth. Upper surface of mandibles with regular rows of subimbricate papillæ as in the other genera. Trophorhinium well-developed. Oncochaetæ slender, always sigmoidal, or with one or more flexures, four, very rarely six to a segment, in two pairs on the dorsal surface of the three thoracic and a variable number of basal abdominal segments; rarely lacking on the pro- and mesothorax. Acrochaetæ long, single or in rows, on the first five or more abdominal segments. Microchaetæ much as in the other genera. While the morphological characters seem to be very constant, the pilosity differs somewhat in the different species, as shown in the following series:

Ps. gracilis and varieties (Text-figs. 3 and 4*a*).—Oncochaetæ occasionally lacking on the pro- and mesothorax, but usually present on all the thoracic and four basal abdominal segments, replaced by simple bristles on the three succeeding segments. Acrochaetæ single on first abdominal, two on second to fifth segments, the more ventral hair smaller. Microchaetæ acute, bristle-like.

Ps. rufomedia.—Acrochaetæ five or six on the side of each abdominal segment, a few smaller homostichous hairs also on the thoracic segments.

Ps. caroli.—Oncochaetæ on all the thoracic and first to fifth abdominal segments. Acrochaetæ single. Microchaetæ very small and delicate so that the integument seems to be very smooth.

Ps. filiformis.—Oncochaetæ lacking on prothorax, rather short and delicate on the meso- and metathorax and basal abdominal segments. Acrochaetæ single, not very stout. Microchaetæ rather long, sparser than in *gracilis*.

Ps. belti var. fulvescens.—Oncochaetæ six on pro- and mesothorax, three on each side of the middorsal line. Acrochaetæ three or four on each side of the abdominal segments, in a transverse row. Microchaetæ sparse as in *filiformis* but coarser and more conspicuous.

Ps. flavidula. Acrochaetæ flagelliform, rather delicate, single. Microchaetæ sparse, distinct.

Ps. sericea var. *fortis.* Oncochaetæ slender, on the three thoracic and first to third abdominal segments. Acrochaetæ stout, single, on the first to sixth abdominal segments. Microchaetæ extremely short and sparse so that the integument appears to be very smooth. Hairs on head scattered and inconspicuous.

Ps. championi var.—Oncochaetæ delicate, present on three thoracic and first to fourth abdominal segments. Acrochaetæ slender, single, on first to sixth abdominal segments. Microchaetæ small, inconspicuous.

Ps. decipiens.—Very similar to *championi,* but the oncochaetæ even more delicate and the acrochaetæ longer and stouter. Antennal rudiments and maxillary sense-organs larger and more heavily chitinized.

Ps. elegans.—Antennal rudiments small. Exudatoria, especially of the pro-thorax, larger than in the other species. Oncochaetæ long and moderately stout, on the three thoracic and first to fifth abdominal segments. Acrochaetæ long, four or five in a regular transverse row on each side of all the abdominal segments. In the young larvæ these hairs are very long and form an uninterrupted transverse row on the ventral and lateral surfaces of each abdominal segment.

Ps. elongata.—Pilosity very much as in gracilis. Acrochaetæ long, delicate, single. Microchaetæ short and sparse.

We may now turn to an analysis of the pellets dissected out of the trophothylaces of the various species of Pseudomyrminæ. As all of these corpuscles (400 in number) contain more or less insect fragments, whereas the latter are frequently absent or very scarce in the infrabuccal pellets of the cospecific workers, we have been led to assume that the corpuscle given to the larva is compounded of accumulated strigil-sweepings (spores, mycelium, pollen, particles of dirt, etc.) together with pieces of freshly captured insect prey. In all probability the adult ant, while malaxating the latter, consumes portions of it before depositing the remainder together with the heterogeneous contents of the infrabuccal pocket in the trophothylax. In other words, the Pseudomyrminæ combine the contents of the dust-bin and garbage-can and serve up the mixture as appropriate food for their young—a truly remarkable example of food-conservation and one certainly very rarely, if ever, exhibited by the ants of the other subfamilies!

While it is rather easy to identify most of the vegetable and inorganic substances in the tropholthylax pellets, this is by no means true of their insect components. The latter comprise small irregular bits of chitin, hairs, pieces of facetted eyes, legs, antennæ, etc. No entomologist is sufficiently expert to refer such fragments even to the order, much less to the family or genus to which they belong, especially when, as in the case of the pellets of the Pseudomyrminæ, he is dealing with fragments of small tropical insects. Only in a few instances, where the fragments were exceptionally large or of insects possessing unique peculiarities of structure (Coccids, Lepidoptera, some Coleoptera) has it been possible to refer them to their family or order.

The species of Pseudomyrminæ are here discussed in the same sequence as in the list on p. 250.

1. *Tetraponera allaborans* (35 pellets).—Practically all of the pellets of this species consist of fragments of small insects, among which pieces of chitin, detached mandibles, and pieces of eyes and hairs can be recognized. Pollen, fungus spores and pieces of mycelium occur in very few instances. Plate IV, Fig. 29 shows pollen grains and numerous small insect fragments, among which portions of the compound eyes of at least two species are clearly discernible.

2. *Viticicola tessmanni* (13 pellets).—The insect substances in the pellets of these larvæ resemble the yolk of ants' eggs and the fat-body of the larvæ themselves, suggesting that some of the brood had been used as food for the more vigorous progeny. In one pellet pieces of the skin of a *Viticicola* larva could be clearly recognized. There are also spores and bits of hyphæ in many cases and particles that seem to be pith and callous tissue. This ant forms very large colonies which nest only, so far as known, in the hollow stems of a singular Verbenaceous liana, *Vitex staudtii*. The pale color of the adult ants, even of the males, the relatively small eyes, the often wingless or sub-apterous females and the peculiar relations of the ants to the host plant as described in another paper (1921?) by the junior author, all indicate a condition of symbiosis more intimate even than that obtaining between certain species of *Pseudomyrma* (*belti, spinicola, canescens*) and the bull-horn acacias. In this connection, the absence from the *Viticicola* pellets of any clearly recognizable insect material obtained outside the myrmecodomatia may be very significant.

3. *Pachysima aethiops* (34 pellets).—Like the preceding, this large black ant is definitely associated with a host plant, in this instance *Barteria fistulosa*, in the inter-nodes of which it lives. It has precisely the same geographical range as the *Barteria* in the western part of the Ethiopian Region. Practically every pellet examined contains pieces of Coccids or the crumpled-up bodies of entire young Coccids. Fungus spores and pieces of mycelium are often abundant as are also pieces of plant-tissue, evidently gnawed from the walls of the cavities (myrmecodomatia) inhabited by the ants. In a few of the pellets the junior author also found small Nematodes resembling the species of *Pelodera* described by Janet (1893*b*, 1893*c*) as living both as parasites in the pharyngeal glands of certain European ants and as free organisms in the

detritus of the nest.   Plate I, Fig. 1 shows an entire mite (?) taken from one of the pellets;   Fig. 3 a cluster of spores mixed with the soft tissues of a Coccid.   Plate II, Fig. 10 shows what is an unmistakable piece of a Coccid, and Fig. 13 an entire speci-men of small size but so perfectly preserved that Prof. Robert Newstead was able to identify it as a larval *Stictococcus formicarius* Newst.   In Fig. 14 there is a collection of spores, plant-hairs and other detritus, and similar materials make up the bulk of the pellet, part of which is shown in Plate V, Fig. 37.

4. *Pachysima latifrons* (2 pellets).—This species is much rarer than the preceding but seems to have very similar habits.   Of the two pellets examined, one contains fragments of Coccids, some spores, bits of mycelium, pollen grains and some pith tissue, with amber-colored cell-contents; the other contains much the same substances together with a few Nematodes.

5. *Pseudomyrma gracilis* (10 pellets).—Half of the pellets are from the larvæ of colonies inhabiting dead twigs, the other half from colonies nesting in the large thorns of acacias.   In the former the pellets consist of bits of insects, often showing the hairs, legs, claws, antennal and tarsal joints and groups of ommatidia, more or less abundant spores, hyphæ and pieces of plant-tissue, especially plant-hairs.   Plate IV, Fig. 28 shows a portion of one of these pellets with its coarse insect fragments. The pellets of the larvæ from acacia thorns are very different in that they lack the bits of insects and contain instead pieces of the food-bodies (Beltian bodies) of the host-plant, mingled with spores and in some instances with a few pollen grains.   The occurrence of the Beltian bodies in the larval food-pellets is interesting, as it shows that *Ps. gracilis*, which nearly always nests in the dead twigs of various trees and bushes, is quite as able, when inhabiting acacias, to utilize the Beltian bodies as food as are the obligatory Pseudomyrmas (*belti*, *spinicola*, *canescens*) of these trees.

6. *Pseudomyrma gracilis* var. *mexicana* (43 pellets).—The long series of pellets of this variety contains an extraordinary collection of the most diverse insect fragments, varying considerably in color and evidently belonging to numerous species.   Some of the pieces are rather large chunks, others finely triturated and are mingled with more or less abundant collections of spores, occasional hyphal fragments and pollen grains.   Among the pollen grains those of pines are easily recognized.   We figure portions of six of the pellets on Plates II to V.   Fig. 11 shows an assortment of spores and pollen grains, Fig. 17 very coarse bits of insects with pollen grains and Fig. 22 a considerable portion of a small crushed (beetle?) larva.   In Fig. 27 the insect frag-ments are more finely divided and mixed with fungus hyphæ, in Fig. 30 much coarser and retaining their bristly hairs.   Fig. 36 represents a small portion of a pellet made up entirely of a great variety of spores, with some insect hairs.

7. *Pseudomyrma gracilis* var. *dimidiata* (29 pellets).—All the pellets contain insect fragments and most of them a variable number and variety of spores and pollen grains. Some of the latter could be recognized as belonging to pines. Of the bits of insects, some are very coarse, as shown in Plate IV, Fig. 31. The fungus spores, of which several are shown in Plate V, Fig. 34, are very peculiar. One of them is represented under a higher magnification in Fig. 32 on the same plate.[1]

8. *Pseudomyrma gracilis* var. nov. (5 pellets).—The components of the pellets of larvæ of this variety taken from large acacia thorns are similar to those of *dimidiata*. Bits of insects, spores and pollen grains, some of which were from pine trees, predominate. Two of the pellets also contain small pieces of plant-tissue. Plate I, Fig. 5 shows large fragments clearly recognizable as portions of a small Curculionid beetle. In trimming the photograph the snout was cut off. Plate IV, Fig. 23 shows a mixture of pollen grains and insect appendages.

9. *Pseudomyrma rufomedia* (21 pellets).—Nearly all the pellets contain numerous bits of insect material, some in chunks. Spores and pollen are also present, being in some cases very abundant, as shown in Plate V, Fig. 33, from a pellet made up entirely of these components.

10. *Pseudomyrma elegans* (2 pellets).—Only two pellets were obtained from a small number of larvæ of this species collected by Mr. Emerson in British Guiana in an earthern termitarium. One contains considerable fragments of a small larva (apparently a Myrmeleonid, judging from a nearly entire eye), a few fungus spores, Lepidopteran scales and pieces of algal filaments. The other, much less voluminous pellet contains bits of an unidentified insect, with small pieces of vegetable tissue (bits of moss leaves?). In neither pellet was there anything that could be identified as pieces of termites or of their eggs.

11. *Pseudomyrma belti* (32 pellets).—In the pellets of this ant, which is an obligatory inhabitant of the thorns of the bull-horn acacias, insect fragments are less abundant than in the various preceding species; spores are also scarce in most cases;

---

[1] These spores, or rather spore-masses, evidently belong to a species of *Ravenelia*, a singular genus of leaf-fungi, the interesting morphology of which was elucidated by Parker many years ago (1886). The specimens were forwarded to Prof. J. C. Arthur, who sent the following comments on them: "I am able to make out that the very interesting rust spores on the slide, which you sent, are those of *R. distans* Arth. & Holw. Not only are the teliospores present but the very characteristic urediniospores also, a half dozen or more of which show in excellent condition. They are unusually small, thin-walled and pointed, with four equatorial pores, making them a very characteristic spore. The host is some Mimosaceous plant, but unfortunately has not determined, as you will see on page 424 of my Uredinales of Guatemala. Only one collection of the rust is known, which came from Retalhuleu. Of course an ant would pick up spores of the rarest fungus possible, and in this case it is not only that the *Ravenelia* is little known, but there are numerous other spores present, which appear to be some species of *Diorchidium*, which I am quite unable to recognize. The genuine Diorchidiums are found on Mimosaceous hosts, but I know of no species from North America and none with spores quite like this."

pollen grains much more abundant. In nearly all cases pieces of Beltian bodies, with occasional bits of other vegetable tissues, are present. Plate II, Fig. 9 shows some of the Beltian bodies, each of which seems to have been merely broken into a few pieces by the workers before being placed in the trophothylax of the larva. Plate V, Fig. 38 shows an abundance of pollen grains, probably derived from the host plants which were in flower when the ants and their larvæ were collected.

12. *Pseudomyrma belti* var. *fulvescens* (21 pellets).—Since the habits of this variety are the same as those of the typical *belti*, it is not surprising to find essentially the same constituents in the trophothylax pellets. There are, however, in addition to the sparse bits of insects, spores, pollen-grains and fragments of Beltian bodies, numerous plant-hairs and bits of miscellaneous vegetable tissues. In a few of the pellets the pollen is largely that of pines. In a series of larvæ taken from acacia thorns in the dry Zacapa region of Guatemala no insect fragments were found and nearly all the pellets contain some pine pollen.

13. *Pseudomyrma spinicola* (1 pellet).—The single pellet obtained from a larva of this Acacia-inhabiting species defies analysis. It consists of a soft, apparently coagulated substance, possibly of vegetable origin, but no cellular elements can be detected in it.

14. *Pseudomyrma sericea* var. *fortis* (6 pellets).—All the pellets taken from the larvæ of this ant, which was found nesting in the internodes of a large palo santo (*Triplaris macombii*), contain numerous large fragments of insects (Plate I, Fig. 7), more or less abundant fungus spores, with bits of hyphæ and of medullary tissue evidently gnawed from the walls of the myrmecodomatia by the workers.

15. *Pseudomyrma championi* var. (2 pellets).—One of the pellets contains a crushed insect (Plate IV, Fig. 25), a 5-toothed mandible of which is left intact, together with numerous spores and pollen grains. The other pellet consists in great part of spores and pollen, with some soft material of unknown origin.

16. *Pseudomyrma filiformis* (9 pellets).—This ant, as the senior author has shown (1919), lives in dead branches in rather shady places. The food of the adults and larvæ appears to be very similar to that of other species of the genus dwelling in dead twigs. The insect fragments are frequently large and coarse, as in the portion of a pellet represented in Plate IV, Fig. 24, and are mingled with fungus spores, often abundant as in this case. Traces of pollen grains and the parenchyma tissue of plants are occasionally present.

17. *Pseudomyrma decipiens* (12 pellets).—Pieces of insects were found in all the pellets, sometimes as more or less hairy bits of chitin, sometimes as chunks or single hairs. Spores are common to very abundant in all but one pellet; pollen grains and

bits of hyphæ are less numerous.   Plate V, Fig. 39 shows a portion of the single pellet referred to, which consists almost entirely of enormous numbers of small spores.

18. *Pseudomyrma caroli.*—No pellets were obtained from the rather small number of larvæ of this species in the collection.

19. *Pseudomyrma elongata* (22 pellets).—In all cases the pellets contain bits of insect material, often in rather large chunks.   In two pellets small entire mites were found.   Fungus spores are more or less abundant in nearly all cases, pollen grains are less numerous and fungus hyphæ very scarce.

20. *Pseudomyrma flavidula* (66 pellets).—In the long series of pellets examined, bits of insects, varying from chunks or whole appendages to minute particles of chitin, scales and hairs are constantly present.   In 35 of the pellets from larvæ collected on Fish Hawk Key, Andros Island, fungus spores, bits of mycelium and pollen are very scarce, in the remaining 31 pellets, from Mangrove Key, Andros Island, spores and bits of hyphæ and miscellaneous vegetable tissue are rather abundant.   Plate III, Fig. 20 shows a pellet fragment made up of soft insect tissues interspersed with fungus spores of many kinds.   In Plate IV, Fig. 26 there are many insect fragments, spores and bits of mycelium.

21. *Pseudomyrma flavidula* var. *delicatula* (22 pellets).—On the whole, the components of the pellets are very similar to those of the typical *flavidula*, but some contain scales of Lepidoptera and in several whole mites were found.   Bacteria could also be recognized among the spores and hyphal elements, which are somewhat less abundant than in the typical form of the species.   Plate I, Fig. 4 shows part of a pellet in which fragments of a small ant can be identified.   Plate II, Fig. 15 shows fungus spores of a singular type mingled with insect-hairs and other débris.

22. *Pseudomyrma arboris-sanctæ.*—Unfortunately no pellets were obtained from the larvæ of this ant, which is an obligatory symbiont of a palo santo (*Triplaris cummingiana*).   Nothing is known of the nature or source of its food.

23. *Pseudomyrma* sp. from Patulul, Guatemala (7 pellets).—All the pellets contain bits of insects, sometimes recognizable as bits of legs, heads of larvæ, eyes, pieces of cuticle, hairs, scales, etc.   The vegetable components, comprising spores, hyphæ, pollen-grains and miscellaneous pieces of tissue, vary from mere traces to large accumulations.   In some pellets there is also a considerable quantity of dirt or very fine detritus, as shown in Plate V, Fig. 35.

24. *Pseudomyrma* sp. from Antigua, Guatemala (2 pellets).—The constitution of these pellets is like that in the preceding species.   In one, fragments of plant epidermis were identified.

25. *Pseudomyrma* sp. from Escuintla, Guatemala (2 pellets).—One pellet contains a few spores, the other fragments of chitin.

26. *Pseudomyrma* sp. from Cartago, Costa Rica (3 pellets).—All the pellets contain bits of chitin, eyes, hairs, etc. and two of them also moderate quantities of spores, bits of mycelium and pollen.

It will be seen that the ingredients of the pellets of the various Pseudomyrminæ as briefly summarized in the foregoing paragraphs, show considerable uniformity in nearly all the species and considerable diversity in the individual pellet. In other words, all the species supply their larvæ with both insect and vegetable substances, but of many different kinds. There can be no doubt that small miscellaneous insects furnish the most important ingredient of the pellets in most species, and that this ingredient, which supplies the most easily assimilable proteids for the growth of the larvæ, is rarely completely lacking even in the acacia-inhabiting species. In the latter the Beltian bodies are unquestionably important sources of food for both the adults and the young. Nor would the spores, hyphæ and pollen grains, which are in most cases at least merely strigil-sweepings, be so constantly fed to the larvæ, unless they could be at least partly utilized as food. That these ingredients, and especially the spores and pollen, contain substances of high nutritive value, is certain, and it is not improbable that the larvæ can triturate them by means of the trophorhinium and thus render them assimilable. This must, indeed, be true, if the spores are actually ingested, for none of them can be detected as whole bodies among the stomach contents as is the case in the above-described Myrmicinæ (*Leptothorax*, *Cryptocerus* and *Cataulacus*), which have no trophorhinium and swallow entire insect fragments and spores.

The constant occurrence of a variety of spores in the infrabuccal pockets of the Pseudomyrminæ is not surprising, when we consider that these very active, large-eyed, wasp-like, arboreal ants, owing to their habit of incessantly patrolling the surfaces of bushes and trees in the tropics, could not fail to accumulate great numbers of the most diverse fungus-germs on their bodies and appendages. The possibility of their behaving as very active and deleterious agents in the spread of many of the fungus-diseases of tropical plants is apparent, but the precise extent of the injury thus caused would depend on whether the workers usually or always feed their infrabuccal pellets to the larvæ or whether, like other ants, they often rid themselves of these corpuscles outside their nests and on the surfaces of plants where the contained spores and bits of hyphæ might germinate. The solution of such problems can, of course, be undertaken only in the tropics.

The trophorhinium, an organ apparently overlooked by previous observers, is beautifully developed in all the Pseudomyrminæ, but seems to show little variation within the subfamily (Text-fig. 4). As stated above, it consists of numerous transverse, parallel, very minutely spinulose ridges in the chitinous cuticle lining the flattened mouth-cavity. If the mouth be carefully opened with the dissecting needles and the dorsal and ventral portions spread apart as in Fig. 5, it will be seen that the dorsal surface or portion (a), corresponding anteriorly to the ventral surface of the

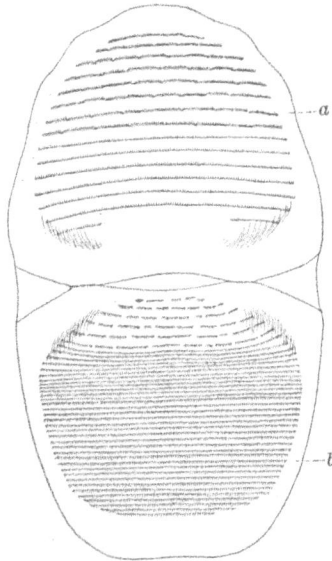

FIG. 5. Trophorhinium of *Pseudomyrma gracilis* Fabr. a, roof. b, floor of mouth.

labrum, begins near its anterior border and extends back nearly to the opening of the gullet. The more anterior ridges are made up of rather arcuate sections, whereas the posterior are straight and more even. The ventral surface or portion (b), corresponding to the floor of the mouth, is of similar structure, except that the ridges are much more numerous and closer together, especially anteriorly. Posteriorly, in the neighborhood of the gullet, they are interrupted and much further apart. On both surfaces the very fine, hair-like spinules point towards the oral orifice. The cuticle of both surfaces is slightly darker than elsewhere on the body. It is not necessary, however, to separate the walls of the buccal cavity as in Text-fig. 5, in order to determine the peculiarities of the two surfaces. This can be done very readily by focussing with the fine adjustment of the microscope on the head of a larva from which all the soft parts have been removed by treatment with caustic potash, or by the study of sections such as the one represented in Text-fig. 6 and 6a. Here the difference in spacing between the fine ridges of the dorsal and ventral surfaces is distinctly seen, the latter beginning just behind the orifice of the salivary duct on the labium and the former extending further forward on the ventral surface of the labrum.

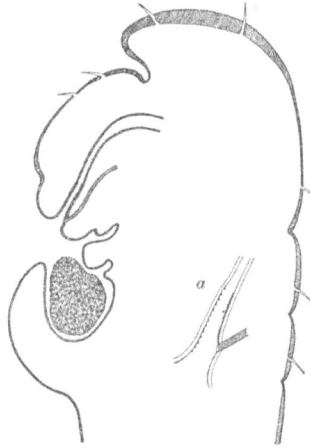

FIG. 6. Sagittal section through anterior end of larva of *Viticicola tessmanni* Stitz. *a*, lining of mouth in section, enlarged.

That an organ of such structure would be admirably adapted to triturating particles of food and sifting or straining out the coarser and harder pieces seems to us to be extremely probable. The process of feeding would appear to be as follows: The pellet placed in the trophothylax by the worker nurse and consisting of the strigil-sweepings, etc. taken from her own infrabuccal pocket *plus* some fragments of insect prey, is probably bathed or saturated with the saliva of the larva secreted into the trophothylax from the orifice of the labial duct. The proteolytic ferment of the secretion would evidently peptonize the softer portions of the particles which could then be drawn back by the mandibles in installments between the two surfaces of the trophorhinium where the indigestible chitinous fragments could be separated out and the remainder turned over to the gullet and swallowed. As a matter of fact, the senior author has seen numerous particles, like those in the trophothylax, spread out between the two apposed surfaces of the trophorhinium in some larvæ that had been suddenly killed by immersion in strong alcohol while apparently in the very act of feeding.

The senior author's study of the trophorhinium in many other ant-larvæ belonging to most of the subfamilies, shows that its delicate transverse ridges have developed from the minutely granular or reticular structure so characteristic of the general chitinous integument. This is beautifully seen in certain Ponerinæ, *e.g.*, in the Ethiopian *Megaponera foetens* Fabr. (Plate III, Fig. 18), where the transverse ridges are still represented by a reticulate areolation like that of the general integument, except that the meshes are drawn out transversely. Among all the ants studied, the trophorhinium seems to attain its most elaborate development in the Ectatommiini among the Ponerinæ and the higher genera of Formicinæ. Fig. 19, Pl. III shows the larval mouth-parts of a species of the former subfamily, the Australian *Rhytidoponera cristata* Mayr. The trophorhinium may be discerned as an arcuate system of extremely fine, parallel ridges extending across the space between the bases of the mandibles. It will be noticed also that the upper surfaces of the latter are very finely and regularly imbricated. This condition likewise obtains, although in some species less conspicuously, in larval Pseudomyrminæ, as stated on p. 256 and as shown in Text-fig. 4. When the mandibles are opened or closed, their imbricated surfaces would naturally rub against the dorsal plate of the trophorhinium, and in forms with large mandibles, like *Rhytidoponera* the latter must overlap more or less when opening and closing and thus also rub against one another. Not improbably, therefore, the rough surfaces of the mandibles may reinforce the triturating and sifting functions of the trophorhinium.

All of these structures, however, seem to have still another function. Comparison of the trophorhinium with the stridulatory organs at the base of the mid-dorsal aspect of the postpetiole and first gastric segment in adult ants of most of the subfamilies (Pseudomyrminæ, Ponerinæ, Dorylinæ and Myrmicinæ) suggests that it may also have a sound-producing function, when it is not being used as a mill or strainer and the two apposed surfaces can be rubbed directly against one another, *i.e.*, without the intervention of food particles. Owing to the small size of the organ and the extreme delicacy of its parallel ridges, the tones produced would be very feeble and of very high pitch, like those produced by the striated surfaces on the bases of the postpetiole and first gastric segment of adult ants. Similarly, the imbricated dorsal surfaces of the mandibles would probably produce faint, shrill sounds when rubbed against one another or against the dorsal surface of the trophorhinium. The resemblance of the latter to the stridulatory organs of the adult becomes even more striking when we recall that the very fine parallel ridges of these organs also arise, as Sharp (1893) and Janet (1893a) have shown, by a simple modification of the general reticular surface-sculpture of the chitinous integument.

That the trophorhinium has the grinding and sorting function we have ascribed to it, is also indicated by certain other facts. The stomach contents of Pseudomyrmine as well as other larvæ possessing a trophorhinium consist of such finely divided particles that their nature cannot be determined, whereas the particles fed to the larvæ (Ponerinæ) or placed in the trophothylax are, as has been shown, very coarse and in most cases easily recognizable as insect or vegetable. On the other hand as stated on p. 255, certain Myrmicine larvæ, like those of *Leptothorax*, *Cryptocerus* and *Cataulacus* have no trophorhinium and swallow their coarse food-particles whole. In fact, the stomachs of these ant-larvæ always contain a collection of insect fragments so large and angular that it is difficult to see how they can pass through the slender gullet. Similarly, the Doryline larvæ have no trophorhinium and take into their stomachs crude though dechitinized pellets of insect tissue.

GENERAL CONSIDERATIONS.

The feeding of the larvæ among ants exhibits a much greater diversity than in any other group of social insects. We were able to distinguish the following methods:

1. Feeding with whole insects or pieces of insects (Ponerinæ, and some Myrmicinæ and Formicinæ);

2. With pellets made of the flesh of insects (Dorylinæ);

3. With the contents of the infrabuccal pocket, either alone or with the addition of fresh insect fragments (Pseudomyrminæ and possibly some Myrmicinæ, such as *Cryptocerus* and *Leptothorax*). In the acacia-inhabiting species of *Pseudomyrma* portions of the Beltian bodies of the host plant are also fed to the larvæ;

4. With pieces of seeds (Granivorous Myrmicinæ);

5. With fungus hyphæ, normal or modified as "kohlrabi," or bromatia. (Tribe Attiini among the Myrmicinæ);

6. With liquids regurgitated from the ingluvies, or crop of the worker (Dolichoderinæ, Formicinæ and many Myrmicinæ).

It is evident that the first method is the most primitive and, owing to the fact that the pieces of insects are often given to the larvæ without malaxation, apparently an even more ancient form of feeding the young than we find in the social wasps. The second method, however, as employed by the Dorylinæ, seems to be very much like that of the higher Vespidæ. All the other methods are highly specialized and are evidently derived secondarily from specializations in the feeding habits of the adults. This is obvious in the granivorous, fungus-growing and honey ants, which represent peculiar adaptations to life in arid or desert environments or to regions in which, during long periods of the year, insect food is very scarce. The conditions in the Pseudomyrminæ are unique, owing to the development in the larvæ of a special post-oral receptacle (trophothylax) for the reception of a food-pellet provided by the worker and consisting of the strigil-sweepings compacted in her infrabuccal pocket *plus* a certain amount of freshly captured and dismembered insect prey. In a study undertaken by the senior author and Mr. George C. Wheeler of the larvæ of a large number of other ant genera, no structure comparable to the Pseudomyrmine trophothylax has been found, except in certain species of *Camponotus* of the subgenus *Colobopsis*. In all the species of the latter subgenus examined the larva is very hypocephalic and the ventral portion of the first abdominal segment projects considerably beyond the thoracic segments and presents a pronounced concavity or basin in the mid-ventral region precisely in the position of the trophothylax of the Pseudomyrminæ. A feeble vestige of the same structure occurs in many *Camponotus* larvæ belonging to other subgenera. No solid pellet is deposited in the basin of *Colobopsis*, but it may, perhaps, be used to hold a supply of the liquid food regurgitated by the workers or of the saliva secreted by the larva itself for the benefit of its attendants.[1]

Some interesting questions are suggested by the composition of the pellet formed in the infrabuccal pocket of imaginal ants. Much of it undoubtedly consists of strigil-sweepings, as Janet observed and as we have found from examination of species of the most diverse genera from widely different stations (deserts, tamarack-bogs, mesophytic and rain-forests, prairies, etc.) and geographical regions. We should, of course, expect insects like the Formicidæ, which are more or less hairy and sculptured, with strongly articulated bodies, nesting in plant cavities or in the ground and constantly moving over the dusty soil and vegetation, to accumulate on the surfaces of their bodies a most heterogenous collection of minute particles and eventually to gather them into their infrabuccal pockets by licking with the tongue or using the strigils. As we have shown, the analysis of the pellets exhibits this diversity very clearly. Many of the components, such as sand-grains, particles of earth and wood, plant and insect hairs, and bits of chitin are inert structures of no further significance, except as they may serve as substrata for the growth of the numerous fungus spores which are such surprisingly constant constituents of the pellets. The junior author has discussed the fungus elements and their possible economic importance in other papers (1920, 1921?). A cultivation of the pellets on artificial media will very probably show that the spores and pieces of hyphæ are quite viable after their sojourn in the infrabuccal pocket. In fact, this is a certainty in the case of the Attiine ants, the recently fecundated queens of which, as von Ihering, J. Huber and Bruch have shown, carry the germs of the fungus gardens of their prospective colonies as pellets in their infrabuccal pockets. The fact that the pellets, even of other species of ants, are cast out by the workers somewhere in their environment, either on the kitchen middens of the nest or outside its precincts, *i.e.*, in situations where the spores may readily germinate, is of no little ecological and economic significance, for it shows that

[1] Cf. the senior author's discussion of the exudates (1918).

ants may be very important agents, or vectors in the distribution of many kinds of fungi in general and of various phytopathogenic fungi in particular. Moggridge (1873) long ago demonstrated the important rôle of the grain-storing ants as distributors of the seeds of the higher plants in arid or desert regions, and Sernander (1906) has shown that many herbaceous plants (myrmecochores) in our northern mesophytic forests are disseminated by some of the common species of *Tetramorium*, *Leptothorax*, *Lasius* and *Myrmica*. Further researches in this field of investigation cannot fail to bring to light many facts of both theoretical and practical value.

We have not had an opportunity to study the mouthparts of larval and imaginal insects of the various orders with a view to determining whether organs similar to the trophorhinium can be detected. It would seem that the fine ridges or striæ on the tongues of many imaginal ants and the delicate rows of minute spinules on their buccal and pharyngeal linings, are analogous structures. Somewhat similar structures seem also to have been seen by Carpenter and MacDowell (1912) and Carpenter (1913) on the maxillulæ of certain Dascyllid larvæ among the Coleoptera and by De Gryse (1915) on the homologous organs of Lepidopteran larvæ. Still more striking is the resemblance of the mouth-lining of spiders, even in certain minute details, to the trophorhinium of ant-larvæ. Lyonet (1832, Pl. 10, Figs. 4 and 20) and Kessler (1849, Pl. 9, Figs. 11 and 12) long ago described the buccal cavity in spiders as flattened, with apposed palatal (dorsal) and lingual (ventral) plates, or surfaces, but did not describe the chitinous structures in sufficient detail. Both surfaces are transversely ridged, the lingual much more finely and densely than the palatal. Moreover, the lingual ridges, at least towards the lateral corners of the buccal cavity, are beset with minute spinules precisely like those on the trophorhinium of ant-larvæ. The whole subject, however, of the finer details in the chitinous lining of the Arthropod mouth, requires a special investigation which would lead far beyond the restricted scope of the present paper.

It was suggested above that the trophorhinium might have a stridulatory function. This was inferred from its peculiar structure and from comparison with the stridulatory surfaces on the middorsal aspect of the postpetiole and first gastric segment in adult ants of the subfamilies Dorylinæ, Ponerinæ, Pseudomyrminæ and Myrmicinæ, as described and figured by Sharp (1893) and Janet (1893, 1900). A study by the senior author and Mr. G. C. Wheeler shows that the trophorhinium is present in the larvæ of all the subfamilies, except the Dorylinæ, Cerapachyinæ and some Myrmicinæ, and that it exhibits in the various tribes and genera numerous, very interesting modifications in detail. A full account of the organ with illustrations is reserved for publication in the near future. A glance at the sketches and photographs of the Pseudomyrmine trophorhinium accompanying this paper (Text-figs. 4, 5 and 6, Plate III, Figs. 18 and 19) shows that the transversely striated dorsal and ventral linings of the larval mouth when rubbed on one another, without intervening particles of food, must produce a tone, extremely faint, to be sure, but not improbably loud enough to be perceived by the worker nurses. And we are not indulging in fancy when we say that this tone may be a hunger cry or cry of distress analogous to that of the human infant.[1]

But in our opinion the trophorhinium is not the only stridulatory organ in ant larvæ. In many species, notably in Ponerinæ, Pseudomyrminæ, some Myrmicinæ (especially the Attiini) and the Formicinæ, the anterior or outer surfaces of the mandibles, as has been stated, are wholly or partly covered with fine, subimbricated papillæ or projections. When the mandibles are merely opened or closed sounds could be produced in two ways, either by a rubbing of their upper surfaces against the dorsal striæ of the buccal lining or by the sharp teeth of one mandible scraping the subimbricated surface of the other. It is highly probable, therefore, that many ant larvæ can produce at least three different sounds, each perhaps associated with a different larval need and capable of eliciting an appropriate response from the attendant workers. If this supposition is correct, the stridulation of the larval brood of a formicary, at least when inadequately supplied with food, would resemble a symphony, or perhaps, more closely, a jazz concert, inaudible to our ears, owing to the very high pitch of its component tones, but perceptible and urgently significant to the workers entrusted with preparing and distributing the larval rations. Perhaps the use of an extremely delicate microphone placed among great masses of large and hungry ant-larvæ may enable us to hear at least some of the lower notes of this shrill chorus. It may also be suggested that the singular habit exhibited by many ants of sorting their larvæ according to size in different parts or chambers of the nest—a habit which has reminded some authors of the division of school-children into classes—may have some connection with stridulation, for if the pitch varies with the size of the larvæ, as there is every reason to suppose, there might be obvious advantages to the nurses in keeping the various stages in groups instead of intermingled.

Although in imaginal insects stridulatory organs are of frequent occurrence they are so rare in larvæ that doubts may arise concerning the accuracy of our interpreta-

---

[1] In this connection the following remarks on the human infant by Mrs. Blanton, quoted by Watson (1919), are interesting: "The cry of one baby can be distinguished with some practice from the cries of another even in a nursery of 25, the overtones varying just as in older people. . . . The 'hunger cry' has generally a well-marked rhythm, the first syllable of preliminary sound coming on the first part of the first beat, the second or accented syllable on the second part of the first beat and a quick intake of breath as the third beat. This measure is most often repeated in groups of 5 or 6, each slightly more forceful than the preceding ones until the fourth or fifth, the last one being softer. Thus also will the groups be repeated. Each measure is also a trifle higher in pitch than the one preceding."

tion of the sound-producing function of the trophorhinium and subimbricated mandibular surfaces. Stridulation is known to occur in some Lepidopteran caterpillars, but according to Prochnow (1912) their sound-producing organs have a very low development "being little different from the sculpture of the integument on other parts of the caterpillars' body," even in *Rhodia fugax* Butl., "which emits a clear and rather loud tone." Concerning the ethological meaning of stridulation in these insects so very little is known that they need not be further considered in this connection. Much more interesting is the case of the large Lamellicorn beetles of the genus *Passalus*, which is widely distributed through the tropics of both hemispheres and is even represented by a species (*P. cornutus*) in the United States as far north as Illinois and Massachusetts. Ohaus (1899–1900, 1909) was able to ascertain from a study of several Brazilian species that these beetles live in rotten logs in families each comprising a male and female with their larval offspring. The senior author has also frequently observed the same composition of these communities in the United States, Central America and Queensland. Ohaus found that the beetles make spacious galleries, comminuting the wood and probably treating the particles with some digestive enzyme (extraintestinal digestion?) so that they can be eaten by the larvæ which follow along the galleries just behind their tunneling parents. Owing to the structure of their mouthparts the larvæ are quite unable to comminute the wood, and when removed from their parents soon die. The beetles not only guard their greenish eggs and diligently provide food for their larvæ but also protect the pupæ and feed the imaginal young till their chitinous integument is completely hardened.

It has long been known that the *Passalus* larva has beautiful stridulatory organs in the form of a broadly elliptical, finely striated area on each of the middle coxæ. The hind leg is reduced to a small, single-jointed appendage, shaped like a mammal's fore paw, with very short digits and claws. The latter are drawn over the striated area as the appendage is worked up and down and produce an audible tone. Schiödte (1862–1873) and Sharp (1899) have published excellent drawings of the organ in our American *Passalus cornutus* and a Bornean species. Plate II, Fig. 12, from a photograph of a Costa Rican species observed by the senior author, is interesting as showing how the fine striæ are made up by a fusion in rows of the minute papillæ of the general integument. We have already called attention to the origin of the stridulatory striæ as modifications of the general integumentary sculpture in the trophorhinia of certain Ponerine ant larvæ and in the postpetiolar and gastric stridulatory areas of many adult ants belonging to the lower subfamilies.

The stridulatory organs of the adult *Passalus* differ greatly in structure and position from those of the larva. The accounts of them given by Leconte (1878),

Ohaus (1900) and Babb (1901) do not, however, agree. Babb clearly describes the organs as a couple of patches of minute denticles on the dorsal surface of the fifth abdominal segment, which rub against specialized areas consisting of ridges and denticles on the ventral surfaces of the folded wings. As in the case of the larva, the tone produced by this apparatus is clearly audible.

The social habits of *Passalus*, as described by Ohaus, clearly suggest that the stridulatory organs of both adults and larvæ must represent means of communication, capable of keeping the members of the family together and in mutual coöperation. That in highly social insects like the ants the even more helpless larvæ should possess elaborate stridulatory organs capable of apprising the attendant workers of such organic needs as those of food, of change of position when the temperature and moisture conditions are unfavorable, of pupation, of the secretion of exudates or of the evacuation of meconium, is not surprising. It would, indeed, be strange if some such method of communication had not developed in so extraordinarily resourceful a group of insects as the Formicidæ.

BIBLIOGRAPHY.

1886. ADLERZ, G. Myrmecologiska Studier II. Svenska Myror och deras Lefnadsförhållanden. *Bihang till K. Svensk. Vet. Ak Akad. Handl.*, 11, 1886, pp. 1–329, 7 pls.

1915. ARNOLD, G. A Monograph of the Formicidæ of South Africa. Part 1. *Ann. S. Afr. Mus.*, 14, 1915, pp. 1–159, 1 pl., 8 text-figs.

1901. BABB, G. F. On the Stridulation of Passalus cornutus Fabr. *Ent. News*, 12, 1901, pp. 279–281, 1 pl.

1920. BAILEY, I. W. Some Relations between Ants and Fungi. *Ecology*, 1, 1920 (in press).

1921(?). BAILEY, I. W. The Anatomy of Certain Plants from the Belgian Congo, with Special Reference to Myrmecophytism. In Wheeler's Report on the Ants collected by Messrs. Lang and Chapin in the Belgian Congo. *Bull. Amer. Mus. Nat. Hist.* (in press).

1841. BRANTS, A. Bijdrage tot de Kennis der Monddeelen van eenige Vliesvleugelige Gekorvenen. (Insecta Hymenoptera.) *Van der Hoeven and de Vries's Tijdschr. voor Natuurlijke Geschiedenis en Physiologie*, 8 Deel., 1841, pp. 71–126, 1 pl.

1919. BRUCH, C. Nidos y Costumbres de Hormigas. *Revist. Soc. Argent. Cienc. Nat.*, 4, 1919, pp. 579–581, 2 figs.

1912. CARPENTER, G. H. AND MABEL C. MACDOWELL. The Mouthparts of some Beetle Larvæ (Dascillidæ and Scarabæidæ). *Quart. Journ. Micr. Sc.*, 57, 1912, pp. 373–396, 3 pls. and 5 text-figs.

1913. CARPENTER, G. H. The Presence of Maxillulæ in Beetle Larvæ. *Trans. 2nd Internat. Congr. Ent.*, Oxford, 1912, pp. 208–215, 2 figs.

1915. DE GRYSE, J. J. Some Modifications of the Hypopharynx in Lepidopterous Larvæ. *Proc. Ent. Soc. Washington*, 17, 1915, pp. 173–178, 3 pls., 1 text-fig.

1915. DONISTHORPE, H. ST. J. K. British Ants, Their Life-history and Classification. Plymouth, England, 1915.

1899a. EMERY, C. Vegetarianisme chez les Fourmis. *Arch. Sc. Phys. Nat.*, Genève (4), 8, pp. 488–490.

1899b. EMERY, C. Intorno alle Larve di Alcune Formiche. *Mem. R. Accad. Sci. Ist.*, Bologna (5), 8, 1899, pp. 3–10, 2 pls.

1912a. EMERY, C.   Alcune Esperienze sulle Formiche granivore.   *Rend. Sess. R. Accad. Sc. Ist.*, Bologna, 1912, pp. 107–117, 1 pl.

1912b. EMERY, C.   Etudes sur les Myrmicinæ.   *Ann. Soc. Ent. Belg.*, 56, 1912, pp. 94–105, 5 figs.

1915.   EMERY, C.   La Vita delle Formiche.   Fratelli Bocca, Torino, 1915.

1906.   ESCHERICH, K.   Die Ameise, Schilderung ihrer Lebensweise.   F. Vieweg u. Sohn, Braunschweig, 1906.

1901.   FIELDE, MISS A. M.   A Study of an Ant.   *Proc. Acad. Nat. Sci. Phila.*, 53, 1901, pp. 425–449.

1905.   HUBER, J.   Ueber die Koloniegründung bei Atta sexdens.   Biol. Centralbl. 25, 1905, pp. 606–619, 625–635, 26 figs.   Transl. in *Smithsonian Report* for 1906, pp. 355–372, 5 pls. 1907.

1898.   IHERING, H. VON.   Die Anlage neuer Colonien und Pilzgärten bei Atta sexdens.   *Zool. Anzeig.*, 21, 1898, pp, 238–245, 1 fig.

1893a. JANET, C.   Etudes sur les Fourmis, les Guêpes et les Abeilles.   Note 1.   Sur la Production de Sons chez les Fourmis et sur les Organes qui les produisent.   *Ann. Soc. Ent. France*, 1893, pp. 159–168.

1893b. JANET, C.   Sur les Nematodes des glandes pharyngiennes des Fourmis (Pelodera sp.).   *C. R. Acad. Sci.*, 117, 1893, pp. 700–702, 1 fig.

1893c. JANET, C.   Etudes sur les Fourmis.   Note 4.   Pelodera des glandes pharyngiennes de Formica rufa.   *Mém. Soc. Zool. France*, 7, 1894, p. 45.

1895a. JANET, C.   Etudes sur les Fourmis.   8ᵐᵉ Note.   Sur l'Organe de Nettoyage tibio-tarsien de Myrmica rubra L. race levinodis Nyl.   *Ann. Soc. Ent. France*, 63, 1895, pp. 691–704, 7 figs.

1895b. JANET, C.   Etudes sur les Fourmis, les Guêpes et les Abeilles.   9ᵐᵉ Note.   Sur Vespa Crabro L. Histoire d'un Nid depuis son Origine *Mém. Soc. Zool. France*, 8, 1895, pp. 1–140, 41 figs.

1897a. JANET, C.   Système Glandulaire tégumentaire de la Myrmica rubra.   Observatio ns diverses sur les Fourmis, Lille, 1897.

1897b. JANET, C.   Sur le Lasius mixtus, l'Antennophorus uhlmanni, etc. Limoges.   1897.

1900.   JANET, C.   Recherches sur l'Anatomie de la Fourmi et Essai sur la Constitution de la Tête de l'Insecte. Carré et Naud, Paris, 1900, 205, pp., 15 pls., 50 text-figs.

1904.   JANET, C.   Observations sur les Fourmis.   Ducourtieux et Gout, Limoges, 1904.

1905.   JANET, C.   Anatomie de la Tête du Lasius niger.   Ducourtieux et Gout, Limoges, 1905, 40 pp., 5 pls. and 2 text-figs.

1849.   KESSLER.   Beitrag zur Naturgeschichte und Anatomie der Gattung Lycosa.   *Bull. Soc. Imp. Natur. Moscow* 22, 1849, pp. 480–523, 1 pl.

1874.   LANDOIS, H.   Thierstimmen.   Freiburg i. Br. 1874.

1878.   LECONTE, J. L.   Stridulation of Coleoptera.   Psyche 2, 1878, p. 126.

1877.   LUBBOCK, SIR JOHN.   On Some Points in the Anatomy of Ants.   *Month. Micr. Journ.*, 18, 1877, pp. 121–142, 4 pls.

1832.   LYONET, P.   Recherches sur l'Anatomie et les Métamorphoses de différentes Espèces d'Insectes. Paris, J. B. Baillière 1832.

1879.   McCOOK, H. C.   The Natural History of the Agricultural Ant of Texas, a Monograph of the Habits, Architecture and Structure of Pogonomyrmex barbatus.   Author's Ed.   *Acad. Nat. Sci. Phila.*, 1879.

1906.   MELANDER, A. L. AND C. T. BRUES.   The Chemical Nature of Some Insects' Secretions.   *Bull. Wis. Nat. Hist. Soc.*, N.S., 4 1906, pp. 22–36.

1860.   MEINERT, F.   Bidrag til de Danske Myrers Naturhistorie.   *Dansk. Vetensk. Selsk.*, 5, 1860, pp. 275–340, 3 pls.

1873.   MOGGRIDGE, J. T.   Harvesting Ants and Trap-door Spiders.   Reeve & Co., London, 1873.

1910.   NEGER, F. W.   Neue Beobachtungen an körnersammelnden Ameisen.   *Biol. Centralbl.*, 30, 1910, pp. 138–150, 3 figs.

1909. NEWELL, W. The Life History of the Argentine Ant. *Journ. Econ. Ent.*, 2, 1909, pp. 174–192, 4 figs., 3 pls.

1899–1900. OHAUS, F. Bericht über eine entomologische Reise nach Zentralbrasilien. *Stettin. Ent. Zeitg.*, 60, 1899, pp. 204–245, 61, 1900, pp. 164–191, 193–274.

1909. OHAUS, F. Bericht über eine entomologische Studienreise in Südamerika. *Stettin. Ent. Zeitg.*, 70, 1909, pp. 3–139.

1886. PARKER, G. H. On the Morphology of Ravenelia glandulæformis. *Proc. Amer. Acad. Arts and Sc.* 22, 1886, pp. 205–219, 2 pls.

1912. PROCHNOW, O. Die Organe der Lautäusserung in Schröder's Handbuch der Entomologie, 1,1912, pp. 61–75, 11 figs.

1862–1873. SCHIÖDTE, J. C. De Metamorphosi Eleutheratorum Observationes: Bidrag til Insekternes Udviklingshistorie. Thieles Bogtrykkeri, Copenhagen 1862–1873.

1906. SERNANDER, R. Entwurf einer Monographie der europäischen Myrmekochoren. *K. Svensk. Vetensk. Handl.*, 41, 1906, pp. 1–410, 11 pls.

1893. SHARP, D. On Stridulation in Ants. *Trans. Ent. Soc. London*, 1893, Part 2, pp. 199–213, 1 pl.

1899. SHARP, D. Insects. Vol. 2. 1899 in Cambridge Natural History. Macmillan & Co., London, 1899.

1914. STITZ, H. Die Ameisen. In Insecten Mitteleuropas insbes. Deutschlands. Bd. 2, 1914.

1917. STRINDBERG, H. Neue Studien über Ameisenembryologie. *Zool. Anzeig.*, 49, 1917, pp. 177–197, 14 figs.

1892. TANNER, J. E. Oecodoma cephalotes. Second paper. Trinidad Field Naturalists' Club, 1, 1892, pp. 123, 127.

1919. WATSON, J. B. Psychology from the Standpoint of a Behaviorist. J. B. Lippincott, Phila., 1919.

1919. WILLIAMS, F. X. Philippine Wasp Studies. *Bull. Hawai. Sugar Planters' Assoc. Ent. Ser.*, 14, 1919, pp. 18–186, 106, figs.

1902. WHEELER, W. M. A New Agricultural Ant from Texas, with Remarks on the Known North American Species. *Amer.* 36, 1902, pp. 85–100, figs.

1904. WHEELER, W. M. A Crustacean-eating Ant (Leptogenys (Lobopelta) elongata Buckley). *Biol. Bull.*, 6, 1904, pp. 251–259, 1 fig.

1907. WHEELER, W. M. On Certain Modified Hairs Peculiar to the Ants of Arid Regions. *Biol. Bull.* 13, 1907, pp. 185–202, 14 figs.

1912. WHEELER, W. M. Observations on the Central American Acacia Ants. *Trans. Second Internat. Ent. Congr.*, Oxford, 1912, pp. 109–139.

1918. WHEELER, W. M. A Study of Some Ant Larvæ, with a Consideration of the Origin and Meaning of the Social Habit among Insects. *Proc. Amer. Phil. Soc.*, 57, 1918, pp. 293–343, 12 figs.

1919. WHEELER, W. M. A Singular Neotropical Ant (Pseudomyrma filiformis Fabricius). *Psyche*, 26, 1919, pp. 124–131, 3 figs.

1920. WHEELER, W. M. The Subfamilies of Formicidæ and Other Taxonomic Notes. *Psyche*, 27, 1920, pp. 46–55, 3 figs.

DESCRIPTION OF PLATES.

PLATE I.

FIG. 1. Portion of larval pellet of *Pachysima aethiops*, showing mite (?) and triturated plant tissue. × 300.

FIG. 2. Strigil, or toilet-organ of *Harpagoxenus sublaevis* Mayr. × 116.

FIG. 3. Portion of larval pellet of *Pachysima aethiops*, showing spores and soft tissue of a Coccid (*Stictococcus formicarius* Newst.) × 180.

FIG. 4. Portion of larval pellet of *Pseudomyrma flavidula* var. *delicatula*, showing fragments of insects (ants). × 58.

FIG. 5. Portion of larval pellet of *Pseudomyrma gracilis* (var. nov.), showing fragments of a weevil. × 78.

FIG. 6. Portion of stomach contents of *Cataulacus egenus* larva, showing spores of different types. × 96.

FIG. 7. Portion of larval pellet of *Pseudomyrma sericea* var. *fortis*, showing fragments of chitin. × 78.

FIG. 8. Portion of stomach contents of *Leptothorax* larva, showing fragments of insects. × 96.

PLATE II.

FIG. 9. Portion of larval pellet of *Pseudomyrma belti*, showing fragments of Beltian food-body. × 58.

FIG. 10. Fragment of Coccid (*Stictococcus formicarius* Newst.) from larval pellet of *Pachysima aethiops*. × 208.

FIG. 11. Portion of larval pellet of *Pseudomyrma gracilis* var. *mexicana* showing pollen and various types of spores. × 330.

FIG. 12. Stridulatory organ of *Passalus* sp. from Costa Rica. × 82.

FIG. 13. Larval *Stictococcus formicarius* Newst. from larval pellet of *Pachysima aethiops*. × 58.

FIG. 14. Portion of larval pellet of *Pachysima aethiops*, showing spores, plant hairs and other detritus. × 330.

FIG. 15. Portion of larval pellet of *Psyeudomyrma flavidula* var. *delicatula*, showing spores of different types. × 330.

FIG. 16. Portion of larval pellet of *Pachysima latifrons*, showing spores, pollen and dirt. × 330.

PLATE III.

FIG. 17. Portion of larval pellet of *Pseudomyrma gracilis* var. *mexicana*, showing fragments of insects and pollen. × 78.

FIG. 18. Mouthparts of larva of *Megaponera foetens* Fabr. showing trophorhinium. × 108.

FIG. 19. Mouthparts of larva of *Rhytidoponera cristata* Mayr showing trophorhinium. × 100.

FIG. 20. Portion of larval pellet of *Pseudomyrma flavidula*, showing soft insect tissue and spores of different types. × 78.

FIG. 21. Portion of larval pellet of *Pseudomyrma flavidula*, showing fragments of mites and spores. × 78.

FIG. 22. Portion of larval pellet of *Pseudomyrma gracilis* var. *mexicana*, showing fragments of larva. × 58.

PLATE I.

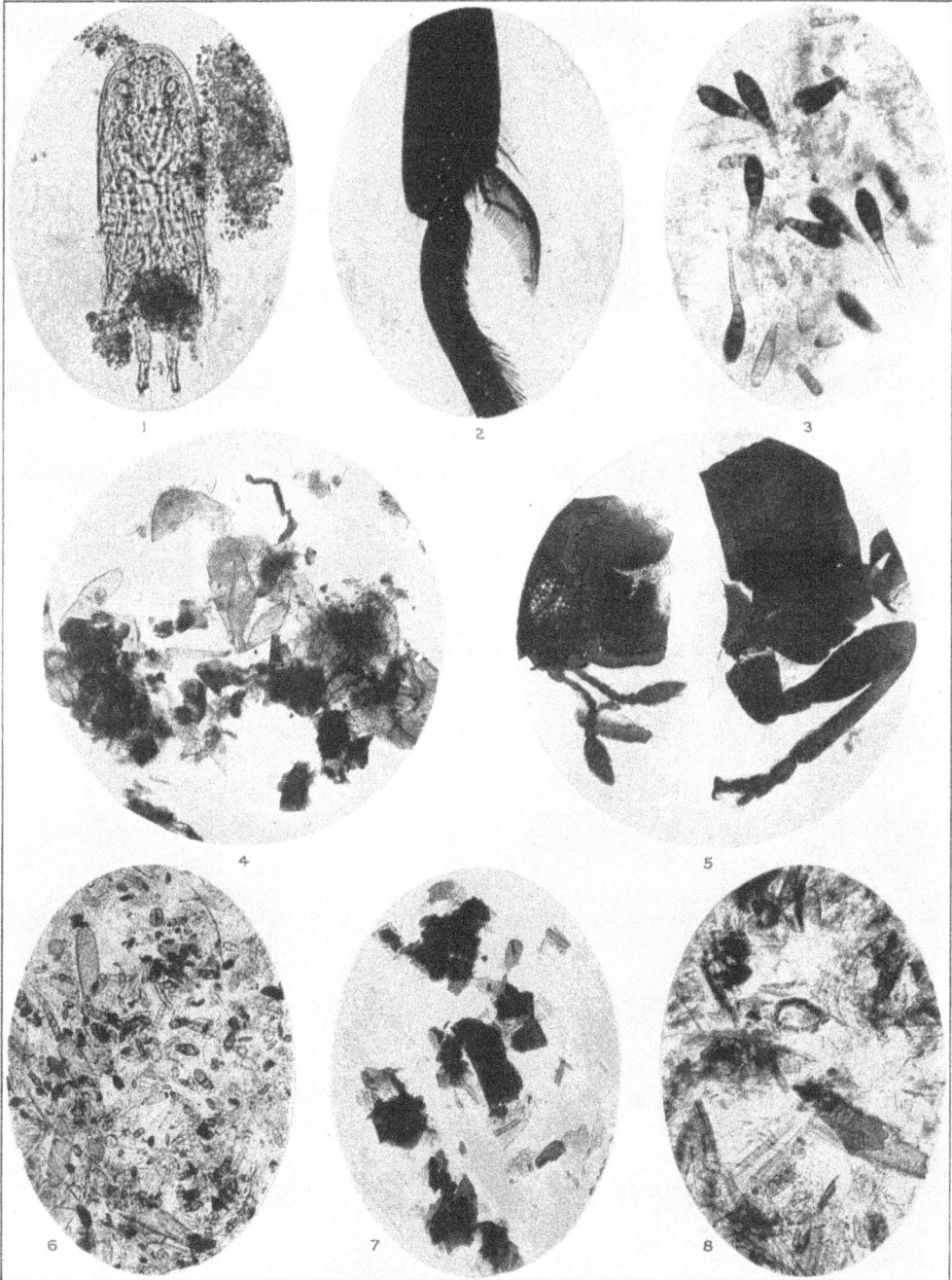

PLATE II.

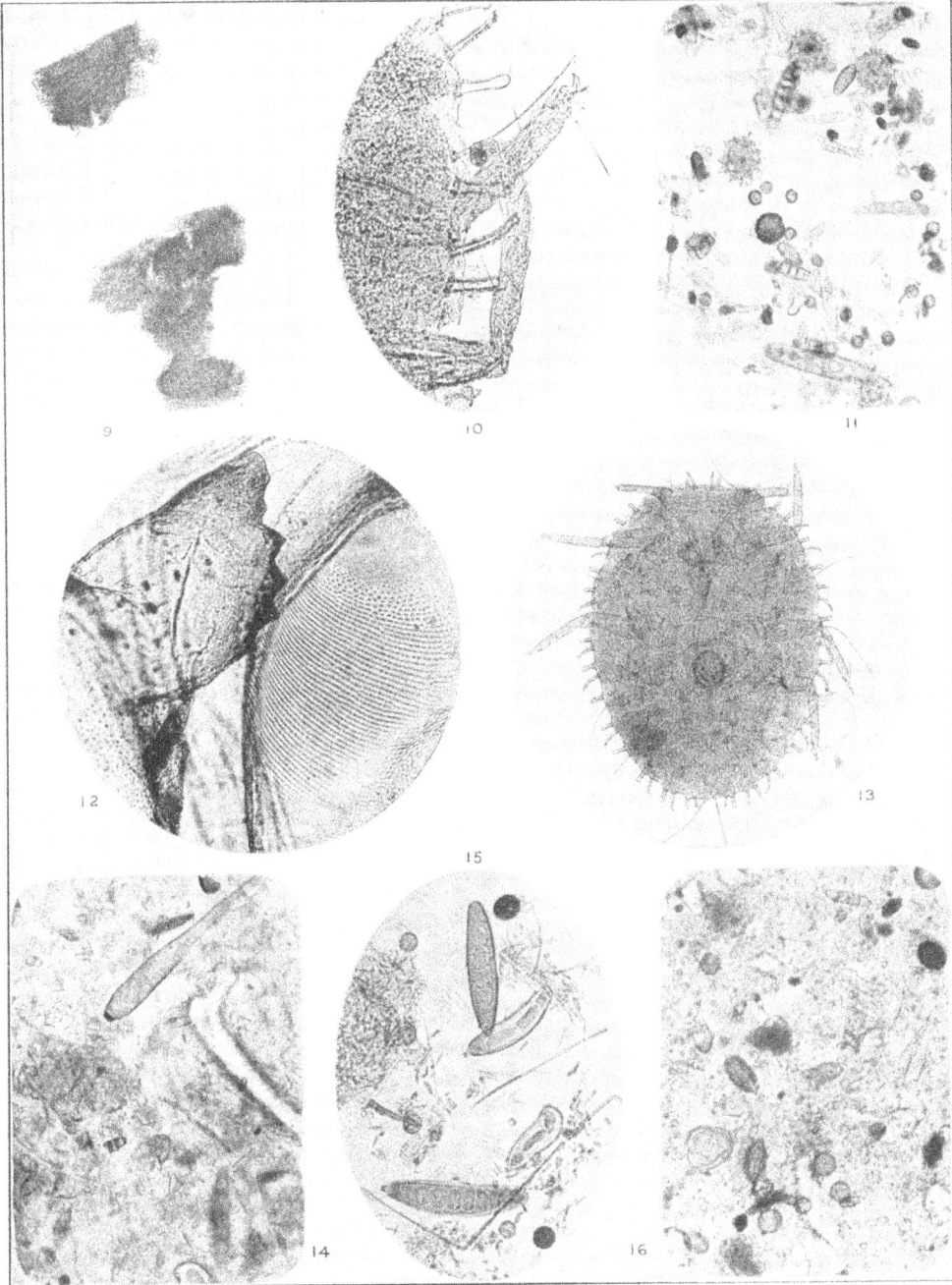

PLATE III.

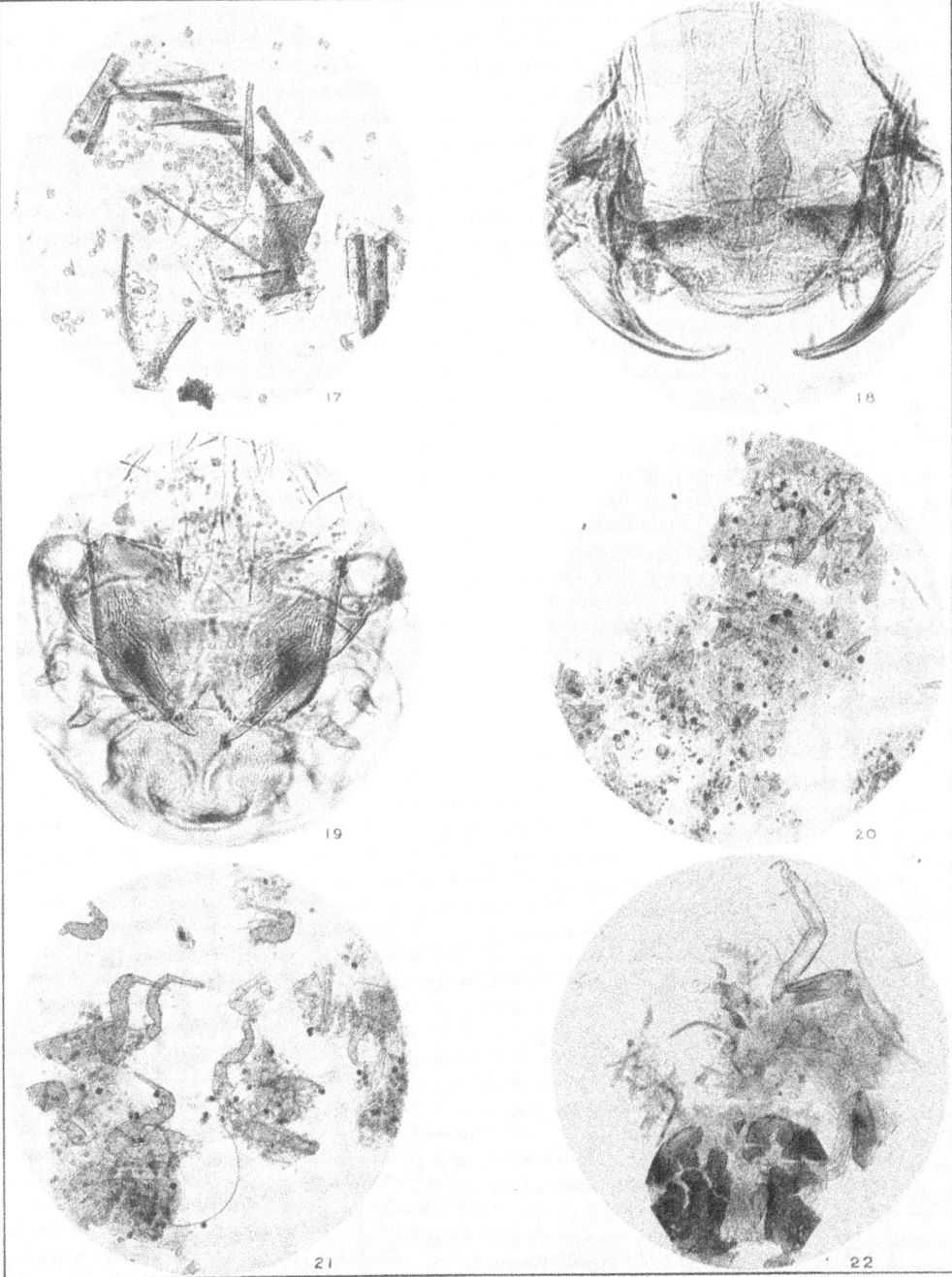

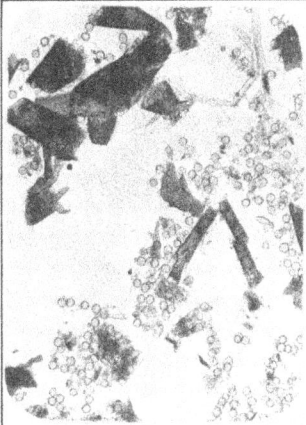

23

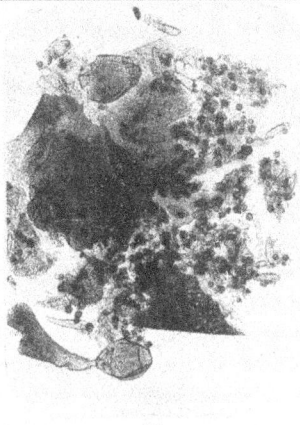

24

25

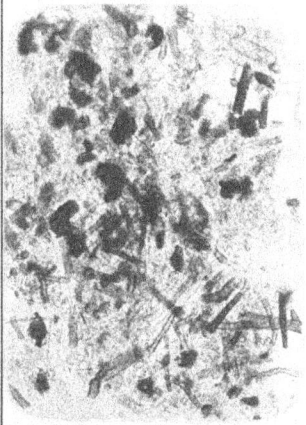

26

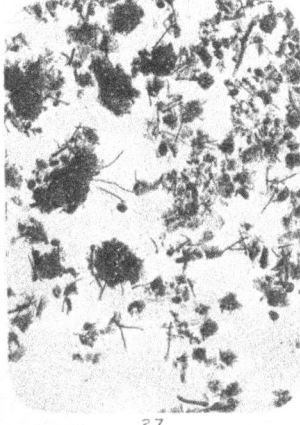

27

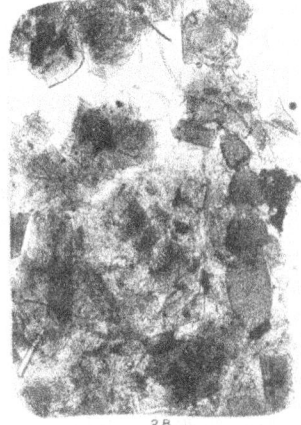

28

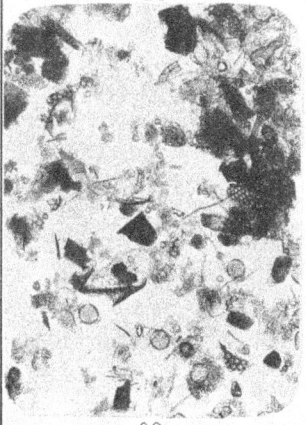

29

30

31

PLATE V.

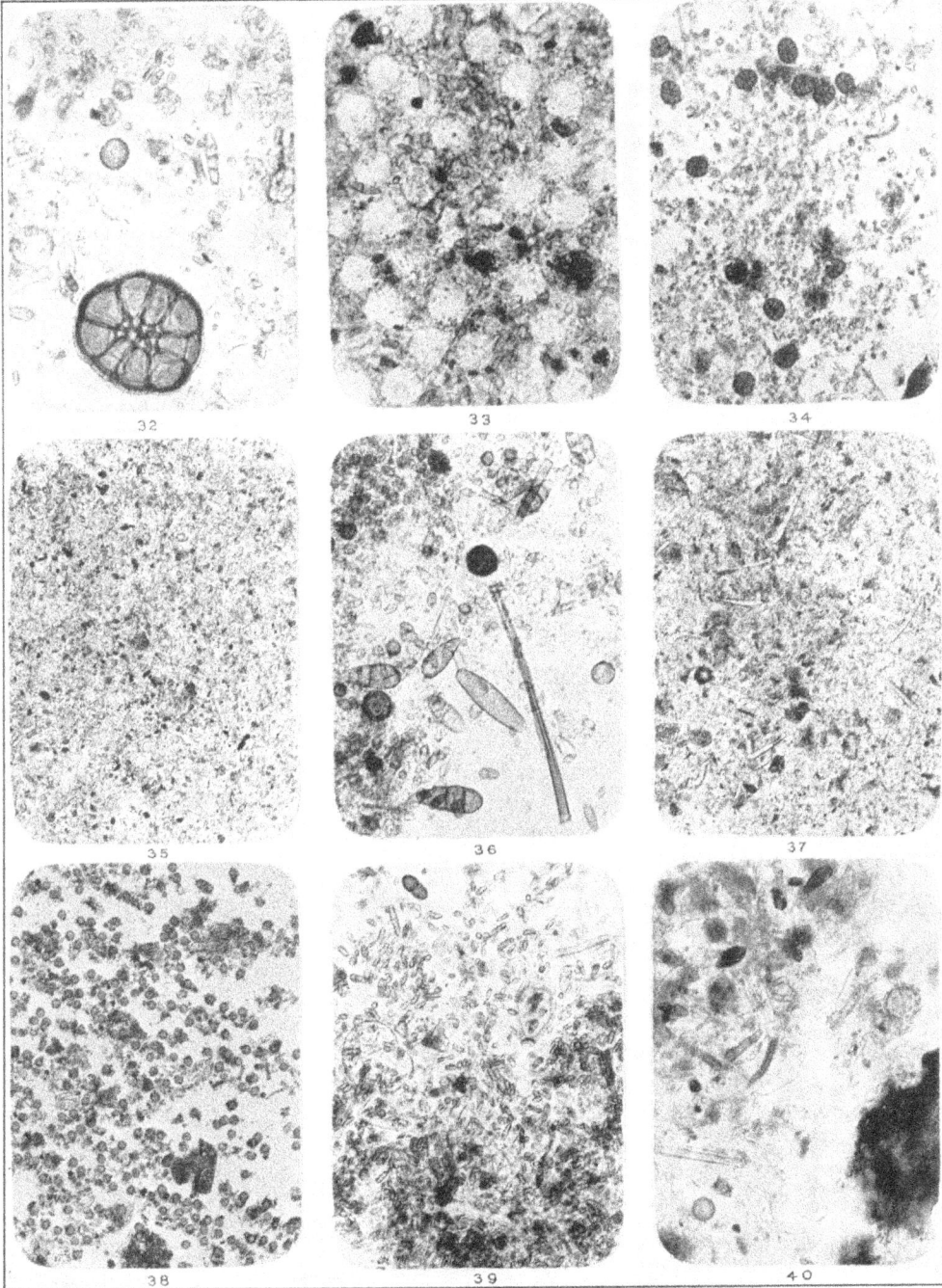

32

33

34

35

36

37

38

39

40

PLATE IV.

Fɪɢ. 23. Portion of larval pellet of *Pseudomyrma gracilis* (var. nov.), showing fragments of insects and pollen. × 58.

Fɪɢ. 24. Portion of larval pellet of *Pseudomyrma filiformis*, showing fragments of insects and spores. × 78.

Fɪɢ. 25. Portion of larval pellet of *Pseudomyrma championi*, showing fragments of insects, spores and pollen. × 78.

Fɪɢ. 26. Portion of larval pellet of *Pseudomyrma flavidula*, showing fragments of insects, hyphæ and spores. × 330.

Fɪɢ. 27. Portion of larval pellet of *Pseudomyrma gracilis* var. *mexicana*, showing fragments of insects and hyphæ. × 58.

Fɪɢ. 28. Portion of larval pellet of *Pseudomyrma gracilis*, showing fragments of insect. × 78.

Fɪɢ. 29. Portion of larval pellet of *Tetraponera allaborans*, showing fragments of insects and pollen. × 78.

Fɪɢ. 30. Portion of larval pellet of *Pseudomyrma gracilis* var. *mexicana*, showing fragments of insect × 78.

Fɪɢ. 31. Portion of larval pellet of *Pseudomyrma gracilis* var. *dimidiata*, showing fragments of insects. × 58.

PLATE V.

Fɪɢ. 32. Portion of larval pellet of *Pseudomyrma gracilis* var. *dimidiata*, showing spores of different types among others those of *Ravenelia distans* A. & H. × 330.

Fɪɢ. 33. Portion of larval pellet of *Pseudomyrma rufomedia*, which is composed entirely of pollen and spores. × 330.

Fɪɢ. 34. Same as Fig. 32. × 58.

Fɪɢ. 35. Portion of larval pellet of *Pseudomyrma* species, from Patulul, Guatemala, showing dirt and other detritus. × 58.

Fɪɢ. 36. Portion of larval pellet of *Pseudomyrma gracilis* var. *mexicana*, which is composed entirely of spores and hairs. × 330.

Fɪɢ. 37. Portion of larval pellet of *Pachysima aethiops*, showing plant hairs, spores and dirt. × 58.

Fɪɢ. 38. Portion of larval pellet of *Pseudomyrma belti*, showing pollen. × 58.

Fɪɢ. 39. Portion of larval pellet of *Pseudomyrma decipiens*, which is composed almost entirely of spores. × 330.

Fɪɢ. 40.   Portion of larval pellet of *Pseudomyrma sericea* var. *fortis*, showing hyphæ and fragments of medullary tissue from myrmecodomatia of *Triplaris macombii*. × 330.

* 9 7 8 1 4 2 2 3 7 7 5 4 3 *